Quantities, Symbols, Units, and Abbreviations in the Life Sciences

Quantities, Symbols, Units, and Abbreviations in the Life Sciences

A Guide for Authors and Editors

Compiled By

Arnošt Kotyk

Institute of Physiology, Academy of Sciences of the Czech Republic, Prague

Humana Press ✳ Totowa, New Jersey

© 1999 Humana Press Inc.
999 Riverview Drive, Suite 208
Totowa, New Jersey 07512

This publication is printed on acid-free paper. ∞
ANSI Z39.48-1984 (American Standards Institute) Permanence of Paper for Printed Library Materials.

Cover design by Patricia F. Cleary.

For additional copies, pricing for bulk purchases, and/or information about other Humana titles, contact Humana at the above address or at any of the following numbers: Tel.: 973-256-1699; Fax: 973-256-8341; E-mail: humana@humanapr.com; Website: http://humanapress.com

Photocopy Authorization Policy:

Printed in the United States of America. 10 9 8 7 6 5 4 3 2 1

Library of Congress Cataloging in Publication Data

Quantities, symbols, units, and abbreviations in the life sciences : a guide for authors and editors /
      compiled by Arnost Kotyk.
          p.   cm.
      Includes bibliographical references (p.   ).
      ISBN 0-89603-616-2 (alk. paper)
      ISBN 0-89603-649-9 (pbk. : alk. paper)
      1. Life sciences—Abbreviations. 2. Life sciences—Notation. I. Kotyk, Arnost.
      QH304.5.Q36   1999   98-31182
      570'.1'48—dc21   CIP

Introduction

In every scientific communication the correct use of symbols, abbreviations, and acronyms for various physical quantities, chemical compounds, biological entities, and so on, has been of paramount importance for ensuring the requisite degree of understanding among scientists from different fields. Quite frequently, scientists do not give much thought to the formal aspects of their publication, as long as the science is sound. Pity the desk editors who labor to make the submitted paper conform to some norm but who, again quite frequently, either do not know the up-to-date norm or create one of their own, different from that of closely related journals.

Quantities, Symbols, Units, and Abbreviations in the Life Sciences: A Guide for Authors and Editors is the result of a four-year discussion among representatives of international scientific unions within the International Council of Scientific Unions (ICSU), various international societies in specialized fields of bioscience, editors of prestigious scientific journals, and eminent individuals active in scientific publishing on the correct use of units, symbols, abbreviations, and acronyms in their particular fields. The initial objective was to collect all the symbols and abbreviations that occur in fields as disparate as biophysics and psychology, to compare them, to see where they differ, and to suggest unification. As the assignment gained momentum, it became clear that: (1) there are more symbols and abbreviations than anticipated, (2) differences in the use of important and sanctioned symbols and abbreviations are not so substantial as originally feared, (3) in some areas there is a notable lack of awareness of what has been accomplished in other quite closely related fields.

All this was revealing and encouraged our work to proceed with a less lofty aim than initially, but with a more practical one, namely, to inform authors, reviewers, and especially desk or copyeditors about the recommendations in different areas of biology and to imbue these persons with the need for some rules and guidelines on the use of "nonwords" in science. Clearly, authors and editors are free to use symbols and abbreviations of their choice as long as they are defined and conform to basic rules of scientific nomenclature in their fields. Authors who disregard the general consensus in this respect deprive their readers of the joy of identifying themselves effortlessly with what the authors wish to convey. They also deprive themselves of the benefit of being cited, except disparagingly.

This guide to the use of units, symbols, abbreviations, and acronyms in biosciences is not a binding, legislative document. It simply brings together the most up-to-date recommendations as they have been promulgated by the appropriate scientific bodies dealing with nomenclatural issues in their respective fields.

Most of the terms defined in *Quantities, Symbols, Units, and Abbreviations in the Life Sciences: A Guide for Authors and Editors* are derived from physics and chemistry and are commonly used in the recommended manner. However, many abbreviations and acronyms created over the past 50 years (and especially many newer ones) are highly arbitrary. This in spite of the fact that some editors recommend their use without definition or explanation. The same is largely true of the form in which they are to be printed (CAPITALS, SMALL CAPITALS, *italics*, **bold-face**, ***boldface italics***, etc.). An attempt is made here to provide guidelines for the creation of such abbreviations or acronyms and to recommend use of a preferred version when alternatives are available.

A few comments on the words used in the title of the book and on some related expressions are necessary.

A "quantity" is here understood in the sense of "a thing that has the property of being measurable in dimensions, amounts, etc., or in extensions of these which can be expressed in numbers and symbols" (Webster's New World Dictionary of American English, Third Edition, 1988), or of "something having magnitude, or size, extent, amount, and the like" (Random House Webster's College Dictionary, 1991; The New Hamlyn Encyclopaedic World Dictionary, 1988). It is interesting that up to the 1950s a "quantity" was defined as what we now describe as "magnitude" (e.g., the unabridged Webster's New International Dictionary, 1948, gives examples of a sphere's surface being a quantity, its area a magnitude, or a yardstick being a quantity, its length a magnitude). This practice is now obsolete and at present length is a quantity, its numerical value its magnitude.

"Symbols" hardly require extensive definition. They are simply letters, Latin or Greek, that represent the quantities.

Quantities are defined, quite rigorously within the SI system, in terms of "units." However, many non-SI units and their names are still in use and are included here (Ångströms, calories, curies, and the like).

"Abbreviations" are either contracted words (concn, wt, yr, and so on) or "initialisms," in which the first letters of words involved (or some prominent letters within the words) are used to form a word that exists independently. "Acronym," now used indiscriminately to describe initialisms of any type (ATP for adenosine triphosphate or EBV for Epstein-Barr virus) was originally restricted to creations that were easy to pronounce as a word (e.g., TRAMP for "tyrosine-rich acidic matrix protein," or SNARE for "soluble *N*-ethylmaleimide-sensitive fusion-protein attachment-protein receptor"). In this book, an abbreviation is viewed as any construct that is shorter than the word(s) it stands for.

The use of contracted words, such as soln for solution, ctrl for control, and the like, is neither commendable nor useful. It represents no appreciable saving of printing space and is reminiscent (1) of certain military documents and (2) of the 17th and 18th century habit of printing (particularly in book or treatise titles) words to fill the space available, often dropping the last one or two letters. Use of these contractions should be kept to an unavoidable minimum (wt for weight, for instance).

In initialisms or acronyms, a preference for three-letter contractions has dominated the field in recent decades—e.g., abbreviations for amino acids, sugars and nucleic acids, for viruses, for physiological factors, diseases, etc. However, many four-letter abbreviations exist, often composed of capital and lowercase letters. There is no reasonable way of attempting a unification here—we simply have to put up with the variety that exists.

An attempt has been made here to list the various abbreviations under separate headings, e.g., Biochemistry and Molecular Biology, Cell Biology, and Animal Physiology. However, it is virtually impossible to draw clear boundaries between many disciplines and there is a great deal of overlap. When abbreviations belong to several areas I have assigned them to the one that appeared first in the sequence of chapters.

The symbols and units listed here are clearly international, even if based mainly on English. The various contracted forms also proceed from the English usage, but are not necessarily the appropriate choice when a text is produced in another language. However, science has become international. This is clearly reflected in that English (either British or American) is the language used for the vast majority of original scientific papers. The "nationalism" that persisted in some major languages exhibited, for instance, by use of RNS for RNA in German, of ARN in French, Spanish, and Italian, and of RNK (PHK) in Russian, is now gone.

Quantities, Symbols, Units, and Abbreviations in the Life Sciences: A Guide for Authors and Editors is neither a terminological dictionary nor a treatise on nomenclature in biology. However, in several cases a definition is given to distinguish one particular symbol or abbreviation from another; in two instances the etymology of descriptive words is given, namely, for the prefixes introduced by the Système International d'Unités and for the chemical elements.

It is the present author's hope that the work will be found useful by a broad range of authors and editors in the biosciences.

In assembling the contents of this book, I made use of several dozen books, booklets, and lists of abbreviations or acronyms. Here are the principal ones.

American College of Clinical Pharmacology, Committee for Pharmacokinetic Nomenclature: Manual of Symbols, Equations and Definitions in Pharmacokinetics. *J. Clin. Pharmacol.* 22:1S-23S (1982).

American Society for Microbiology: ASM Style Manual for Journals and Books. *ASM*, Washington, 1991.

Atomic Weights of the Elements 1991. *Pure Appl. Chem.* 64, 1519–1534 (1992).

Bailar J.C., Mosteller F.: Medical Uses of Statistics. *NEJM Books*, Boston, 1992.

Bassingthwaighte J.B., Chinard F.P., Crone C., Goresky C.A., Lassen N.A., Reneman R.S., Zierler K.L.: Terminology for mass transport and exchange. *Am. J. Physiol.* 250:H539–H545.

Clutterbuck A.: Gene symbols in *Aspergillus nidulans*. *Genet. Res.* 21:291–296.

Cohen E.R., Giacomo P.: Symbols, Units, Nomenclature and Fundamental Constants in Physics. *Physica* 146A, 1–68 (1987).

Committee on Standardized Genetic Nomenclature for Mice: Rules for nomenclature of chromosome anomalies. In: Genetic Variants and Strains of the Laboratory Mouse (Lyon M.F., Searle A.G., eds), pp. 574-575, *Oxford Univ. Press*, New York, 1989.

Council of Biology Editors Style Manual. *American Institute of Biological Sciences*, Washington, 1995.

de Bolster M.W.G.: Glossary of Terms Used in Inorganic Biochemistry. *International Union of Pure and Applied Chemistry* (draft as of February 1995).

Demerec M., Adelberg E.A., Clark A.J., Hartman P.E.: A proposal for a uniform nomenclature in bacterial genetics. *Genetics* 54:61–76 (1966).

Dupayrat J.: Dictionary of Biomedical Acronyms and Abbreviations, 2nd edition. *John Wiley and Sons*, Chichester, 1990.

Ford C., Pollock D., Gustavsson E.: Proceedings of the First International Conference for the Standardisation of Banded Karyotypes of Domestic Animals. *Hereditas* 92:145–162 (1980).

Genetic Nomenclature Guide. Supplement to *Trends in Genetics*. Elsevier, 1995.

Guidelines for Soil Description. *Food and Agriculture Organization of United Nations*, Rome, 1990.

Hawksworth D.L.: A Draft Glossary of Terms Used in Bionomenclature. *International Union of Biological Sciences*, Paris, 1994.

Herbert W.J., Wilkinson P.C., Scott D.I: The Dictionary of Immunology. *Academic Press*, London, 1995.

International Organization for Standardization: ISO Standards Handbook 2: Quantities and Units. *ISO*, Geneva, 1993.

International Standards Organization. International standard ISO 8601: 1988(E): Data elements and interchange of formats-information exchange-representation of dates. *ISO*, Geneva, 1988.

International Union of Biochemistry and Molecular Biology, Nomenclature Committee: Biochemical Nomenclature and Related Documents. *Portland Press*, London, 1992.

International Union of Pure and Applied Chemistry, Applied Chemistry Division, Committee on Biotechnology: Selection of terms, symbols and units related to microbial processes: Recommendations. *Pure Appl. Chem.* 64:1047–1053 (1992).

International Union of Pure and Applied Chemistry, Macromolecular Division, Commission on Macromolecular Nomenclature: Compendium of macromolecular nomenclature. *Blackwell Scientific*, Oxford (1991).

Kidd K.K., Bowcock A.M., Pearson P.L., Schmidtke J., Willard H.F., Track R.K., Ricciuti F.: Report of the Committee of Human Genome Mapping by Recombinant DNA Techniques. *Cytogenet. Cell Genet.* 49:132–218 (1988).

Lackie J.M., Dow J.A.T.: The Dictionary of Cell Biology, 2nd edition. *Academic Press,* London, 1995.

Le Système International d'Unités. *Bureau International des Poids et Mesures,* Sèvres, 1991.

Lentner C.: Geigy Scientific Tables, Vol. 5, Heart and Circulation. *Ciba-Geigy*, West Caldwell, NJ, 1990.

Lindsley D.L., Zimm G.G.: The genome of *Drosophila melanogaster. Academic Press*, San Diego, CA (1992).

Mills I., Cvita T., Homann K., Kallay N., Kuchitsu K.: Quantities, Units and Symbols in Physical Chemistry. *Blackwell Sci. Publ.* Oxford, 1993.

Murphy F.A., Fauquet C.M., Bishop D.H.L., Ghabrial S.A., Jarvis A.W., Martelli G.P., Mayo M.A., Summers M.D. (eds): Classification and nomenclature of viruses: Sixth report of the International Committee on Taxonomy of Viruses. *Springer-Verlag*, New York, 1994.

Oxford Dictionary of Biochemistry and Molecular Biology, *Oxford University Press*, Oxford, 1997.

Perkins D., Radford A., Newmeyer D., Björkman M.: Chromosomal loci of *Neurospora crassa. Microbiol. Rev.* 46:426–570 (1982).

Price C.A.: Nomenclature of sequenced plant genes. *Plant Mol. Biol. Reporter* 12 (supplement), S1–S109 (1994).

Quantities, Units, and Symbols. *The Royal Society*, London, 1975.

Recommendations on Terms and Symbols in Pharmacology. *International Union of Pharmacology* (draft as of January 1994).

Rigg J.C., Visser B.F., Lehmann H.P.: Nomenclature of derived quantities. *Pure Appl. Chem.* 63, 1307–1311 (1991).

Salisbury F.B.: Units, Symbols, and Terminology for Plant Physiology. *International Association for Plant Physiology* (draft as of September 1994).

Scientific Style and Format: The CBE Manual for Authors, Editors and Publishers. *Cambridge University Press*, Cambridge, UK, 1994.

Sherman F.: Genetic nomenclature. In: Molecular Biology of the Yeast *Saccharomyces cerevisiae* (Strathern J., Jones E., Broach J, eds), pp. 639–640. *Cold Spring Harbor Laboratory*, Cold Spring Harbor, 1981.

Shows T.B., McAlpine P.J., Boucheix C., Collins F.S., Conneally P.M., Frezal J., Gershowitz H., Goodfellow P.N., Hall J.G., Issitt P. and others: Guidelines for human gene nomenclature: an international system for human gene nomenclature (ISGN 1987). *Cytogenet. Cell Genet.* 46:11–30.

Singleton P., Sainsbury D.: Dictionary of Microbiology and Molecular Biology. *John Wiley and Sons*, Chichester, 1993.

Skerman V.B.D., McGowan V., Sneath P.H.A.: Approved lists of bacterial names. *Int. J. Syst. Bacteriol.* 30:225–420 (1980).

Sneath P.H.A., Sokal R.R.: Numerical Taxonomy. *W.H. Freeman*, San Francisco, 1973.

Standing Committee of Human Cytogenetic Nomenclature: An international system for human cytogenetic nomenclature (ISCN 1978). *Cytogenet. Cell Genet.* 14:313–404 (1978).

Stedman's Concise Medical and Allied Health Dictionary. *Williams & Wilkins*, Baltimore, 1997.

Stenesh J.: Dictionary of Biochemistry and Molecular Biology. *John Wiley and Sons*, New York, 1989.

The Chicago Manual of Style, 14th edition. *The University of Chicago Press*, Chicago, 1993.

Leading journals in genetics, physiology, immunology, microbiology, and paleontology were consulted as to the use of special symbols and abbreviations.

My personal thanks are due to friends and colleagues from various international scientific bodies and particularly scientific unions within ICSU:

W.E.H. Blum, Vienna, A	soil science
G. den Boef, Amsterdam, NL	physical chemistry
M.W.G. de Bolster, Amsterdam, NL	inorganic biochemistry
G.J. Cosmides, Bethesda, MD, USA	pharmacology
M. Ito, Wako, J	physiology
H. Metzger, Bethesda, MD, USA	immunology
A.T.C. North, Leeds, GB	biophysics
K. Pawlik, Hamburg, G (aided by B. Overmier)	psychology
T. Younès, Paris, F	general biology

Arnošt Kotyk

Scientific Bodies Associated with the *International Council of Scientific Unions* (ICSU)

Unions

IAU	International Astronomical Union
IBRO	International Brain Research Organization
IGU	International Geographical Union
IMU	International Mathematical Union
ISSS	International Society of Soil Science
IUAES	International Union of Anthropological and Ethnological Studies
IUBMB	International Union of Biochemistry and Molecular Biology
IUBS	International Union of Biological Sciences
IUCr	International Union of Crystallography
IUFoST	International Union of Food Science and Technology
IUGG	International Union of Geodesy and Geophysics
IUGS	International Union of Geological Sciences
IUHPS	International Union of the History and Philosophy of Science
IUIS	International Union of Immunological Sciences
IUMS	International Union of Microbiological Societies
IUNS	International Union of Nutritional Sciences
IUPAB	International Union for Pure and Applied Biophysics
IUPAC	International Union of Pure and Applied Chemistry
IUPAP	International Union of Pure and Applied Physics
IUPHAR	International Union of Pharmacology
IUPS	International Union of Physiological Sciences
IUPsyS	International Union of Psychological Science
IUTAM	International Union of Theoretical and Applied Mathematics
IUTOX	International Union of Toxicology
URSI	Union Radio Scientifique Internationale

Committees

CCBS	Committee on Capacity Building in Science
CODATA	Committee on Data for Science and Technology
COMSCEE	Special Committee on Science in Central and Eastern Europe and the Former Soviet Union
COSPAR	Committee on Space Research
SCAR	Scientific Committee of Antarctic Research
SC-IDNDR	Special Committee for the International Decade for Natural Disaster Reduction
SC-IGBP	Scientific Committee for the International Geosphere Biosphere Programme
SCOPE	Scientific Committee on Problems of the Environment
SCOR	Scientific Committee on Oceanic Research
SCOSTEP	Scientific Committee on Solar-Terrestrial Physics
SCOWAR	Scientific Committee on Water Research

CONTENTS

INTRODUCTION v
    Scientific Bodies Associated with the *International Council of Scientific Unions* (ICSU) xi

MATHEMATICS 1
    Dimensionless quantities 2
    Arabic and Roman numerals 2

STATISTICS 3
    Greek alphabet 4

PHYSICS 5
    Base units 5
    Space and time 7
    Classical mechanics 8
    Electricity and magnetism 9
    Radiation 10
    Some useful physical and mathematical constants 11
    Units and magnitudes of logarithmic quantities 12

CHEMISTRY AND BIOLOGY 13
    Right subscripts 13
    Left subscripts 13
    Right superscripts 13
    Left superscripts 14

GENERAL CHEMISTRY 15
    Aggregation states 16
    Chemical elements 16

PHYSICAL CHEMISTRY 31
ORGANIC AND STRUCTURAL CHEMISTRY 35
BIOCHEMISTRY AND MOLECULAR BIOLOGY 39
    Amino acids 39
    Sugars and derivatives 40

Nucleic acids, their components and derivatives 41
Plant lectins 44
Vitamins 44
Miscellaneous abbreviations in biochemistry 44
Miscellaneous abbreviations in molecular biology 48
Methods and techniques in biochemistry and biophysics 50
Reagents and technical terms related to methods 54
Buffers 56
Enzyme kinetics 57
Restriction endonucleases 58

CELL BIOLOGY 63
GENETICS 65
Chromosomes 65
Genes and their expression products 67
Prokaryotes 67
Eukaryotic microorganisms 68
Plants 70
Animals 70

TAXONOMY 73
Suffixes for taxonomical categories 74

VIROLOGY 77
MICROBIOLOGY 83
PLANT PHYSIOLOGY 85
ANIMAL PHYSIOLOGY 87
PHARMACOLOGY 91
IMMUNOLOGY 95
MEDICINE 101
General 101
Diseases 103
Syndromes 112
Techniques 117

SOIL BIOLOGY 119
Soil phases 119
Soil moisture regime 119
Soil temperature regime 119

Vegetation 119

Soil color 120

Soil texture 120

Human-introduced crops 120

Master soil horizons 120

PALEONTOLOGY 121

PSYCHOLOGY 123

Units uniquely used 123

Acronyms for tests, examinations, etc. 123

Acronyms for terms 124

NOTES ON ABBREVIATIONS IN GENERAL USE IN SCIENTIFIC WRITING 129

Quantities, Symbols, Units, and Abbreviations in the Life Sciences

Mathematics

The way numerals and algebraic symbols are written must be respected in all scientific publishing, including biosciences, and therefore the basic rules are given here, but not the abbreviations and acronyms specific to mathematics as a scientific discipline.

As in all subsequent recommendations, preference is given to symbols that are simpler to print and are unambiguous.

Symbol	Meaning	Symbol	Meaning
$=$	equal to	$<$	less than
$\neq$	not equal to	$>$	greater than
$\equiv$	identical with	$\leq$	less than or equal to
$\approx$	approximately equal to	$\geq$	greater than or equal to
$\cong$	asymptotically equal to	$<<$	much less than
$\sim$	proportional to	$>>$	much greater than
∞	infinity	$+$	plus
$-$	minus	$\times$ or $\cdot$	times, multiplied by
$\pm$	plus or minus	$\div$ or $/$	divided by
		(In European usage, a : sign is often used instead.)	

$ab, a\,b, a \cdot b, a \times b$	a multiplied by b
$a/b, ab^{-1}, \dfrac{a}{b}$ (not recommended)	a divided by b
$\lvert a \rvert$	magnitude, absolute value of a
$[a]$	mean value of a
a^n	a to the power of n
$(a + b)^{1/n}$	nth root of $(a + b)$; preferred to $\sqrt[n]{(a + b)}$

Σ	sum	Π	product
$\sin x$	sine of x	$\cos x$	cosine of x
$\tan x$	tangent of x	$\cot x$	cotangent of x
$\sec x$	secant of x	$\operatorname{cosec} x$	cosecant of x

$\arcsin x, \arccos x$, etc.	inverse functions of x (preferred to $\sin^{-1}x$, $\cos^{-1}x$, etc.)
$\sinh x, \cosh x$, etc.	hyperbolic functions of x
$\operatorname{arsinh} x, \operatorname{arcosh} x$, etc.	inverse hyperbolic functions of x
$n!$	factorial of n

(continued)

1

e	base of natural logarithms
e^x, exp x	exponential of x
ln x, $\log_e x$	natural (Napierian) logarithm of x
$\log_a x$	logarithm to base a of x
$\log_{10} x$, lg x	logarithm to base 10 of x
$\log_2 x$, lb x	logarithm to base 2 of x
$\lim_{x \to a} y$	limit of y as x approaches a
i	square root of -1
df/dx, f'	first derivative of f
$d^n f/dx^n$, f^n	nth derivative of f
$\partial f/\partial x$	partial derivative of f with respect to x
$\int f(x)dx$	integral of f
$\boldsymbol{a}$	vector a
grad V, ∇V	gradient of scalar field V
div A, ∇A	divergence of vector field A
A	matrix of elements A_{ij}

Dimensionless Quantities

%	percent = 10^{-2}
‰	permille = 10^{-3}
ppm	parts per million = 10^{-6}
ppb	parts per billion = 10^{-9}

(With the exception of %, the symbols should be avoided; even the use of % should be minimized.)

Arabic and Roman Numerals

0		12	XII	60	LX
1	I	13	XIII	70	LXX
2	II	14	XIV	80	LXXX
3	III	15	XV	90	XC
4	IV	16	XVI	100	C
5	V	17	XVII	500	D
6	VI	18	XVIII	1000	M
7	VIII	19	XIX	5000	$\overline{V}$
8	VIII	20	XX	10 000	$\overline{X}$
9	IX	30	XXX	100 000	$\overline{C}$
10	X	40	XL	1 000 000	$\overline{M}$
11	XI	50	L		

Statistics

Only the commonly occurring symbols are given here. Designations for special techniques in statistics, such as CRD for *completely randomized design* or LSDT for *least significant difference test,* are outside the scope of this book, and there does not seem to exist a generally accepted recommendation for their use.

ANOVA	analysis of variance
CV	coefficient of variation
f	frequency, number of items occurring in a category
F	variance ratio
H_0	null hypothesis
j	index of measurements within a sample in ANOVA
k	number of categories compared by the chi-square test
M	median
n, N	total number of individuals
p, P	level of significance (probability of wrongly rejecting the null hypothesis)
R	coefficient of multiple correlation
r	sample correlation coefficient
r^2	coefficient of determination
s, SD, SD	standard deviation of the sample
$s_{\bar{x}}$, SEM, SEM	standard error of the mean (use of SD or SD and SEM or SEM is discouraged, except perhaps in table headings)
s^2	sample variance
t	statistical value derived from Student's *t*-test
$\bar{x}, \overline{X}$	arithmetic mean of the sample
β	regression coefficient of population
μ	mean of the population
ρ	population correlation coefficient
σ	standard deviation of the population
σ^2	variance of the population
χ^2	statistical value derived in the chi-square test

Greek Alphabet

A α	alpha	H η	eta	N ν	nu	T τ	tau
B β	beta	Θ θ,ϑ	theta	Ξ ξ	xi	Y υ	upsilon
Γ γ	gamma	I ι	iota	O o	omicron	Φ ϕ,φ	phi
Δ δ,∂	delta	K κ	kappa	Π π	pi	X χ	chi
E ε	epsilon	L λ	lambda	P ρ	rho	Ψ ψ	psi
Z ζ	zeta	M μ	mu	Σ σ,ς	sigma	Ω ω	omega

Physics

In what follows we shall observe the recommendations, now almost universally accepted, called the Système International d'Unités, as introduced and codified by The International Organization for Standardization (ISO) as the institution that governs, on a worldwide scale, the use of verbal designations for physical quantities, their symbols and units.

Following a chaotic period when, even in science, the metric system was used together with the British–American usage of inches and pounds, there was a time when the so-called **cgs** (centimeter–gram–second) system was introduced and taught in schools for perhaps 50 years. The system, although adequate for many purposes, was intrinsically nonconvertible without introducing numerical coefficients. It also contained units that were derived independently and rather arbitrarily, such as *calorie* as the amount of heat (energy) required to raise the temperature of 1 gram of distilled water from 3.5 to 4.5 °C (*small calorie*), or from 14.5 to 15.5 °C (*normal calorie*), or, finally, as 1/100th of the heat required to raise its temperature from 0 to 100 °C (*mean calorie*).

It was through the use of the meter–kilogram–second (**mks**) system, which was self-consistent, that an internationally accepted version was introduced in the early 1960s, the SI system. This system is now recommended by all serious scientific bodies in the world and its fundamentals are reported here.

There are seven so-called base units, which are mutually irreplaceable and which can be used in various combinations to describe all the phenomena of the physical world.

Base Units

Name	Symbol	Quantity	Name	Symbol	Quantity
meter	m	length	kelvin	K	temperature
kilogram	kg	mass	mole	mol	amount of substance
second	s	time	candela	cd	luminous intensity
ampere	A	electric current			

The base units are defined as follows:

The *meter* is the length of path traveled by light in vacuum during a time interval of 1/299 792 458 of a second.

The *kilogram* is the mass of the international prototype created by the IIIrd General Conference on Weights and Measures in 1901.

The *second* is the duration of 9 192 631 770 periods of the radiation corresponding to the transition between the two hyperfine levels of the ground state of cesium-137.

The *ampere* is the current that, if maintained in two straight parallel conductors of infinite length, of negligible circular cross section, placed 1 meter apart in vacuum, will produce a force between these conductors equal to 0.2 micronewton per meter.

The *kelvin* is the fraction 1/273.16 of the thermodynamic temperature of the triple point of water (= 0.01 °C).

The *mole* is the amount of substance containing as many elementary entities (atoms, molecules, ions, electrons, photons, etc.) as there are atoms in 12 grams of carbon-12.

The *candela* is the luminous intensity, in a given direction, of a source that emits monochromatic radiation of frequency 0.54 petahertz and that has a radiant intensity in that direction of 1/683 watt per steradian.

Supplementary units, *radian* (rad) for plane angle, and *steradian* (sr) for solid angle, both dimensionless, can be used if appropriate.

Multiples and submultiples of these units (and any other SI-derived units) are

10^{-24}	yocto	y	10^1	deca	da
10^{-21}	zepto	z	10^2	hecto	h
10^{-18}	atto	a	10^3	kilo	k
10^{-15}	femto	f	10^6	mega	M
10^{-12}	pico	p	10^9	giga	G
10^{-9}	nano	n	10^{12}	tera	T
10^{-6}	micro	μ	10^{15}	peta	P
10^{-3}	milli	m	10^{18}	exa	E
10^{-2}	centi	c	10^{21}	zetta	Z
10^{-1}	deci	d	10^{24}	yotta	Y

The etymology of the prefixes is shown below.

yocto   similar to Latin *octo* (eight)
zepto   similar to Latin *septem* (seven)
atto   from Danish *atten* (eighteen)
femto   from Danish *femten* (fifteen)

pico	from Spanish *pico* (tip or peak) or from Italian *piccolo* (small)
nano	from Greek *nanos* (dwarf)
micro	from Greek *mikros* (small)
milli	from Latin *mille* (thousand)
centi	from Latin *centum* (hundred)
deci	from Latin *decimus* (tenth)
deka	from Greek *deka* (ten)
hecto	from Greek *hekaton* (hundred)
kilo	from Greek *chilioi* (thousand)
mega	from Greek *megas* (large)
giga	from Greek *gigas* (giant)
tera	from Greek *teras* (monster)
peta	similar to Greek *pente* (five)
exa	similar to Greek *hex* (six)
zetta	similar to Latin *septem* (seven)
yotta	similar to Latin *octo* (eight)

Most of the quantities occurring in physics are listed here, together with the recommended symbols and units, both special, where they exist, and SI-derived ones.

The fields that are not included are those that have no direct relevance for biosciences, such as atomic physics, quantum mechanics, and the like.

Again no acronyms or abbreviations used for the entities of interest in purely physical research are shown here. There is, however, a list of acronyms for various methods and techniques that are in fact physical at the end of the section on chemistry.

Space and Time

Name	Symbol	Units
length	l	m
height	h	m
thickness	d, δ	m
distance	d	m
path length	s	m
radius	r	m
diameter	d	m

[In structural chemistry and crystallography the ångström (Å) still survives, 1 Å being equal to 0.1 nm.]

area	A, A_s, S	m^2
volume	V	m^3

[A common unit here is the *liter*, equivalent of dm^3, abbreviated L or l; although not derived from a person's name, capital L should be used to avoid confusion of lower-case l with the numeral 1.]

(continued)

Space and Time (continued)

Name	Symbol	Units
molar volume	V_m	$m^3 mol^{-1}$
plane angle	$\alpha, \beta, \gamma, \theta, \phi$	1 (rad)
solid angle	Ω, ω	1 (sr)
time, period	t	s

[Commonly used units here are *min* for *minute* (= 60 s), *h* (not *hr*) for *hour* (= 3600 s), and *d* for *day* (= 86 400 s); the symbol *a* for *year* (*annus*) should be used sparingly, e.g., in geochronological tables. Also permissible but not recommended are *mo* for *month* and *yr* for *year*.]

frequency	v, f	Hz (hertz), s^{-1}
angular frequency	$\omega = 2\pi v$	$s^{-1} (rad\ s^{-1})$
angular velocity	$\omega = d\emptyset /dt$	$s^{-1}\ (rad\ s^{-1})$
damping coefficient	δ	1
relaxation time, time constant	τ, T	s
velocity (a vector)	$\boldsymbol{v, u, w, c}$	$m\ s^{-1}$
speed (a scalar)	v, u, w, c	$m\ s^{-1}$

[It is unfortunate that most fonts available in the text editors contain an italic v or v that rather resembles the Greek letter v (nu).]

acceleration	a	$m\ s^{-2}$
sedimentation coefficient	$s = v/a$	s

[If the sedimentation coefficient is applied to assessing the molar mass of macromolecules, it is expressed in svedbergs (S), 1 *svedberg* being equal to 10^{-13} s.]

Classical Mechanics

Name	Symbol	Units
mass	m	kg
reduced mass	$\mu = m_A m_B/(m_A + m_B)$	kg
density	$\rho = m/V$	$kg\ m^{-3}$
surface density	$\rho A, \rho S = m/A$	$kg\ m^{-2}$
specific volume	$v = V/m = 1/\rho$	$m^3 kg^{-1}$
momentum	$\boldsymbol{p} = m\boldsymbol{v}$	$kg\ m\ s^{-1}$
moment of inertia	$I, J = \Sigma m_i r_i^2$	$kg\ m^{-2}$
force	$\boldsymbol{F} = d\boldsymbol{p}/dt = m\boldsymbol{a}$	N (newton), $m\ kg\ s^{-2}$

[In common practice an F is used instead of a boldface $\boldsymbol{F}$.]

energy	E	J (joule), $m^2\ kg\ s^{-2}$
- , potential	$E_p = -\int F\ ds$	J
- , kinetic	$E_k = \frac{1}{2} mv^2$	J
work	$W, w = \int F ds$	J
pressure	$p, P = F/A$	Pa (pascal), $m^{-1}\ kg\ s^{-2}$
lateral pressure	$\pi = F/d$	$N\ m^{-1}$

(continued)

Classical Mechanics (continued)

Name	Symbol	Units
surface tension	$\gamma, \sigma = (dW/dA)$	N m^{-1}, J m^{-2}
surface pressure	$\pi^s = \gamma^0 - \gamma$	N m^{-1}
[π^0 is the surface tension of a clean surface.]		
weight	$G = m\, g_n$	N
normal stress	$\sigma = F/A$	Pa
shear stress	$\tau = F/A$	Pa
linear strain	$\varepsilon = \Delta l/l$	1
modulus of elasticity	$E = \sigma/\varepsilon$	Pa
shear strain	$\gamma = \Delta x/d$	1
viscosity	$\eta = \tau_{x,z}/[dv_x/dz]$	Pa s
kinematic viscosity	$v = \eta/\rho$	m^2 s^{-1}
fluidity	$\o = 1/\eta$	Pa^{-1} s^{-1}
power	$P = dW/dt$	W (watt), J s^{-1}, m^2 kg s^{-3}

Electricity and Magnetism

Name	Symbol	Units
electric charge	Q	C (coulomb), A s
charge density	$\rho = Q/V$	C m^{-3}
(inner) electric potential	$V, \o = dW/dQ$	V (volt), J C^{-1}, m^{2}kg s^{-3}A^{-1}
(outer) electric potential	$\psi = Q/4\pi\varepsilon_0 r$	V
electric potential difference	$U, \Delta\o = V_1 - V_2$	V
electromotive force	$E = \int(F/Q)\,ds$	V
electric displacement	$\boldsymbol{D} = \varepsilon\boldsymbol{E}$	C m^{-2}
electric field strength	$\boldsymbol{E} = \boldsymbol{F}/Q$	V m^{-1}
capacitance	$C = Q/U$	F (farad), C V^{-1}, m^{-2} kg^{-1} s^4 A^2
permittivity	$\varepsilon = \boldsymbol{D}/\boldsymbol{E}$	F m^{-1}
relative permittivity	$\varepsilon_r = \varepsilon/\varepsilon_0$	1
[This was formerly called *dielectric constant*.]		
electric dipole moment	$\boldsymbol{p} = \sum Q_i \boldsymbol{r}_i$	C m
electric current density	$\boldsymbol{j}\, (I = \int \boldsymbol{j}\, dA)$	A m^{-2}
electric current	$I = dQ/dt$	A
magnetic flux density	$\boldsymbol{B} = \boldsymbol{F}/Qv$	T (tesla), V s m^{-2}, kg s^{-2}A^{-1}
[This is equivalent to *magnetic field*.]		
magnetic flux	$\Phi = \int \boldsymbol{B}dA$	Wb (weber), V s, m^2 kg s^{-2}A^{-1}
magnetic field strength	$\boldsymbol{H} = \boldsymbol{B}/\mu$	A m^{-1}
permeability	$\mu = \boldsymbol{B}/\boldsymbol{H}$	N A^{-2}, H m^{-1}
electric resistance	$R = U/I$	Ω (ohm), V A^{-1}, m^2 kg s^{-3} A^{-2}

(continued)

Electricity and Magnetism (continued)

Name	Symbol	Units
conductance	$G = 1/R$	S (siemens), Ω^{-1}, m^{-2} kg^{-1} s^3 A^2
resistivity	$\rho = E/j$	Ω m
conductivity	$\kappa = 1/\rho$	S m^{-1}
(self-)inductance	$L = E/(dI\,/dt)$	H (henry), V A^{-1} s, m^2 kg s^{-2} A^{-2}

Radiation

Name	Symbol	Units
wavelength	λ	m
wavenumber	$\sigma = 1/\lambda$	m^{-1}
frequency	$v = c/\lambda$	Hz
angular frequency, pulsatance	$\omega = 2\pi v$	s^{-1}, rad s^{-1}
luminous flux	lm (lumen)	cd sr
illuminance	lx (lux)	cd sr m^{-2}
luminous intensity	I_v	cd
radiant energy	Q	J
radiant intensity	I	W sr^{-1}
radiant energy density	$\rho = Q/V$	J m^{-3}
radiant power	$\Phi = dQ/dt$	W
photon intensity	I_p	s^{-1} sr^{-1}
irradiance (radiant flux received)	$I = d\Phi\,/dA$	W m^{-2}
fluence	$F = \int I\,dt$	J m^{-2}
transmittance	$\tau,\ T = I_{tr}/I_0$	1
absorbance, decadic	$A = -\log_{10} T$	1

[Formerly called *extinction* or *optical density*. The latter term is still used when dealing with opaque emulsions or suspensions, such as in microbiology; however, *attenuance* (*D*) is suggested for conditions when scattering is substantial.]

absorptance, absorptivity	$a = I_{abs}/I_0$	1
reflectance	$\rho = I_{refl}/I_0$	1
absorption coefficient		
-, decadic	$a = A/l$	m^{-1}
-, molar	$\varepsilon = a/c$	m^2 mol^{-1}
refractive index	$n = c_0/c$	1
refractive index measured with light of the sodium *D*-lines	n_D	1

(continued)

Radiation (continued)

Name	Symbol	Units
absorption index	$k = 2.3\ a/4\pi\nu$	1
molar refraction, refractivity	$R_m = V_m(n^2 - 1)/(n^2 + 2)$	$m^3\ mol^{-1}$
specific optical rotatory power	$[\alpha]_{\lambda}^{\theta} = \alpha/\gamma\lambda$	$rad\ m^2\ kg^{-1}$
quantum yield of fluorescence	Y_F	1
lifetime of fluorescence	τ_F	s
radiant intensity of a beam resolved into directions parallel and perpendicular to the direction of polarization of the exciting radiation	$\Phi_{\parallel}, \Phi_{\perp}$	$W\ sr^{-1}$
degree of polarization	$P = (\Phi_{\parallel} - \Phi_{\perp})/(\Phi_{\parallel} + \Phi_{\perp})$	1
degree of anisotropy	$r = (\Phi_{\parallel} - \Phi_{\perp})/(\Phi_{\parallel} + 2\Phi_{\perp})$	1
(radio)activity		Bq (becquerel), s^{-1}

[The old unit *curie* (Ci) equal to 37 GBq is still occasionally used; likewise, *counts per min* (cpm) or *disintegrations per min* (dpm) frequently occur in printouts from radioactivity counters with some justification, dpm being defined as cpm corrected for quenching, geometric factors, and the like.]

absorbed dose (of radiation)	D	Gy (gray), $J\ kg^{-1}$, m^2s^{-2}
dose equivalent		Sv (sievert), $J\ kg^{-1}$, m^2s^{-2}

Some Useful Physical and Mathematical Constants

speed of light *in vacuo*	c_0	299 792 458 m s^{-1} (by definition)
permeability of vacuum	$V\mu_0$	$4\pi \cdot 10^7$ H m^{-1} (by definition)
permittivity of vacuum	$\varepsilon_0 = 1/\mu_0 c_0^2$	$8.854\ 187\ 816 \cdot 10^{-12}$ F m^{-1}
Planck's constant	h	$6.626\ 075\ 5 \cdot 10^{-34}$ J s
elementary charge	e	$1.602\ 177\ 733 \cdot 10^{-19}$ C

[From this, 1 electronvolt (eV) is equal to 1.602 177 733 · 10^{-19} J.]

atomic mass constant (unified atomic mass unit)	$m_u = u$	$1.660\ 540\ 2 \cdot 10^{-27}$ kg
Avogadro constant	L, N_A	$6.022\ 136\ 7 \cdot 10^{23}\ mol^{-1}$
Loschmidt's number	N_L	$2.686763 \cdot 10^{25}\ m^{-3}$
Boltzmann's constant	k, k_B	$1.380\ 658 \cdot 10^{-23}$ J K^{-1}
Faraday's constant	$F = eL$	$9.648\ 530\ 9 \cdot 10^4$ C mol^{-1}
gas constant	R	$8.314\ 510$ J $K^{-1}mol^{-1}$
zero of Celsius scale		273.15 K (by definition)
molar volume (at $p = 10^5$ Pa; $t = 0$ °C)		22.711 08 L mol^{-1}

standard atmosphere	atm	101 325 Pa
gravitational constant	G	$6.6873 \pm 0.0094 \cdot 10^{-11}$ $m^3kg^{-1}s^{-2}$
standard acceleration of free fall	g_n	9.806 65 m s^{-2} (by definition)
ratio of circle's circumference to its diameter	π	3.141 592 653 59

Units and Magnitudes of Logarithmic Quantities

base of natural logarithms	e	2.718 281 828 46
natural logarithm of 10	ln 10	2.302 585 092 99
neper (Np)		damping coefficient $\delta \times$ time t
radian (rad)		angular frequency $\omega \times$ time t
decibel (dB)		$10 \log_{10} (P/P_0)$ where P_0 is the reference power (by convention equal to 100 pW m^{-2} for a tone of 1 kHz) and P is the measured power

Chemistry and Biology

This section on symbols in chemistry opens with a list of some of the subscripts and superscripts used. They should be in roman (upright) type, in general, but italicized if they refer to quantities, whose symbols are themselves written in italics.

Right Subscripts

I, II, ..., 1, 2	different systems, different states
1, 2, 3,...	number of atoms per entity
A, B,...	molecular species A, B, etc.
i	typical ionic species
u	undissociated molecule
p, V, T	at constant pressure, volume, temperature
p, m, c, a	with equilibrium constant, in terms of pressure, amount of substance, concentration, activity
g, l, s, c	referring to gas, liquid, solid, crystalline phase
f, e, s, t, d	referring to fusion, evaporation, sublimation, transition, dissolution or dilution
c	critical state or value
∞	limiting value at infinite dilution

Left Subscripts

1, 2, 3,...	are only used for atomic numbers of nuclides

Right Superscripts

⊖,°	standard

[The *plimsoll* ⊖ sign may not be available in the printer used; the symbol for *degree* ° is acceptable.]

*	pure substance; excited electronic state
id	ideal
E	excess
+	positive charge
−	negative charge

13

...$^{-II}$, $^{-I}$, ^{0}, ^{I}, ^{II}...   oxidation number

·   radical

[The radical sign should precede the charge sign, thus $O_2^{\cdot-}$.]

‡   activated complex

Left Superscripts

^{1}, ^{2}, ^{3}...   are only used for mass numbers of nuclides

General Chemistry

Name	Symbol	Units
number of entities	N	1
amount of substance, chemical amount		
(of substance D)	$n_D = N_D/L$	mol
atomic mass	m_a, m	kg
relative atomic mass	A_r	1
mass of entity	m_f, m	kg
atomic mass constant (mass unit)	$m_u = u = m_a(^{12}C)/12$	kg

[The unit commonly used in biochemistry is the *dalton* (Da); for practical purposes, the dalton is equal to 1 g/mol.]

molar mass (of substance D)	$M_D = m/n_D$	kg mol^{-1}
relative molar (molecular) mass	$M_r = m_f/m_u$	1

[The use of <u>molecular weight</u> for this quantity is discouraged.]

specific volume	$v = V/m$	m^2

[All "specific" quantities are understood as the principal quantity divided by mass.]

molar volume (of substance D)	$\overline{V}_{m,D} = V/n_D$	m^{3}mol^{-1}
partial molar volume (of substance D)	$\overline{V}_{m,D} = (\partial V/\partial n_D)_{p,T}$	m^{3}mol^{-1}
mass fraction	$w_j = m_j / \Sigma m_i$	1
volume fraction	$\varnothing_j = V_j / \Sigma V_i$	1
mole fraction (of substance D)	$x_j = n_j / \Sigma n_i$	1
mass concentration (density)	$\gamma, \rho = m/V$	kg m^{-3}
number concentration (of substance D)	$C_D, n_D = N_D/V$	m^{-3}, L^{-1}
amount concentration (of substance D)	$c_D, c(D), [D] = n_D/V$	mol m^{-3}

[This quantity can be expressed as mol dm^{-3} or, quite appropriately, as mol L^{-1}; since this is a clumsy unit, particularly in adjectival constructions, such as *5 mmol/L sucrose*, the use of either M or M (for molarity) is generally accepted but the use of *M*, as occurs particularly in American journals, should be discouraged because it is the proper symbol for molar mass. *Normality* (N) for the "concentration" of equivalents, is not acceptable.]

molality (of solute D)	$m_D, b_D = n_D/m_{solvent}$	mol kg^{-1}

[The m_D symbol is unfortunate as it is identical with that for <u>mass</u>, b being preferable; however, the word <u>molal</u> (prefixed if convenient) should be used instead of any symbol.]

solubility (of solute D)	$s_D = c_D$ (saturated)	mol m^{-3}
surface concentration (of solute D)	$\Gamma_D = n_D/A$	mol m^{-2}
stoichiometric number	v	1
(negative for reactants, positive for products)		
extent of reaction, advancement	$\xi = (n_D - n_{D,0})/v_D$	mol

Aggregation States (always in parentheses after the symbol)

gas or vapor	g	vitreous	vit
liquid	l	amorphous	am
solid	s	adsorbed substance	a, ads
condensed phase (s or l)	cd	monomeric	mon
fluid phase (g or l)	fl	polymeric	pol
crystalline	cr	solution	sln
liquid crystal	lc	aqueous solution	aq

Chemical Elements

Atomic number	Symbol	Relative atomic mass	Name *English* *German* *French* *Russian*	Etymology **E** **G** **F** **R**
1	H	1.008	hydrogen Wasserstoff hydrogène водород	Greek: *hydor* = water + *genes* = born = water substance like E = water-born

(D is *deuterium* with relative atomic mass of 2.0140.)
(T is *tritium* with relative atomic mass of 3.01605.)

2	He	4.003	helium Helium hélium гелий	Greek: *helios* = sun
3	Li	6.941	lithium Lithium lithium литий	Greek: *lithos* = stone
4	Be	9.012	beryllium Beryllium béryllium бериллий	Latin: *beryllus* = beryl
5	B	10.811	boron Bor bore бор	medieval Latin: *borax*, from Persian *būrah*
6	C	12.011	carbon Kohlenstoff carbone углерод	Latin: *carbo* = charcoal = coal substance like E = born of coal

Chemical Elements (continued)

Atomic number	Symbol	Relative atomic mass	Name English German French Russian	Etymology E G F R
7	N	14.007	nitrogen	Greek: *nitron* = native soda + *genes* = born
			Stickstoff	= choking substance
			azote	Greek: *a* = without + *zoe* = life
			азот	like F
8	O	15.999	oxygen	Greek: *oxys* = acid + *genes* = born
			Sauerstoff	= acid substance
			oxygène	like E
			кислерод	= born of acid
9	F	18.998	fluorine	Latin: *fluere* = to flow
			Fluor	
			fluor	
			фтор	
10	Ne	20.180	neon	Greek: *neos* = new
			Neon	
			néon	
			неон	
11	Na	22.990	sodium	from *soda*
			Natrium	Arabic *naṭrūn* = soda
			sodium	like E
			натрии	like G
12	Mg	24.305	magnesium	*Magnesia*, town in Turkey (present-day Manisa)
			Magnesium	
			magnésium	
			магний	
13	Al	26.982	alumin(i)um	Latin: *alumen* = alum
			Aluminium	
			aluminium	
			алюминий	
14	Si	28.086	silicon	Latin: *silex* = flint
			Silizium	
			silicium	
			кремний	кремень = flint

(continued)

Chemical Elements (continued)

Atomic number	Symbol	Relative atomic mass	Name English German French Russian	Etymology E G F R
15	P	30.974	phosphorus Phosphor phosphore фосфор	Greek: *fosforos* = bringer of light
16	S	32.066	sulfur Schwefel soufre сера	Latin: *sulphur* (from Indo-European root *suelplos*)
17	Cl	35.453	chlorine Chlor chlore хлор	Greek: *chloros* = green
18	Ar	39.948	argon Argon argon аргон	Greek: *argos* = inert, idle
19	K	39.098	potassium Kalium potassium калий	Dutch: *potasch* = potash Medieval Latin, from Arabic: *al qali* = saltwort ashes like E like G
20	Ca	40.078	calcium Kalzium calcium кальций	Latin: *calx* = lime
21	Sc	44.956	scandium Scandium scandium сандий	Medieval Latin: *Scandia* = Scandinavia
22	Ti	47.88	titanium Titan titane титан	Greek: *Titan*, a giant son of Uranus and Gaea

Chemical Elements (continued)

Atomic number	Symbol	Relative atomic mass	Name *English* *German* *French* *Russian*	Etymology E G F R
23	V	50.942	vanadium Vanad vanadium ванадий	Old Norse: *Vanadis* = Freya, goddess of love and beauty
24	Cr	51.996	chromium Chrom chrome хром	Greek: *chroma* = color
25	Mn	54.938	manganese Mangan manganèse марганец	Italian: *manganese* (from Magnesia; *see* magnesium)
26	Fe	55.847	iron Eisen fer железо	Indo-European: *eis* = strong like E Latin: *ferrum* = iron *cf.* Greek *chelys* = turtle
27	Co	58.933	cobalt Kobalt cobalt кобальт	German: *Kobold* = goblin
28	Ni	58.34	nickel Nickel nickel никель	contraction of Swedish *kopparnickel* = false, devilish copper
29	Cu	63.546	copper Kupfer cuivre медь	Greek: *kyprios* = Cyprus Greek: *Media* (region in present-day Iran), or Hittite: *miti* = red
30	Zn	65.39	zinc Zink zinc цинк	German: *Zinke* = (originally) prong

(continued)

Chemical Elements (continued)

Atomic number	Symbol	Relative atomic mass	Name English German French Russian	Etymology E G F R
31	Ga	69.723	gallium Gallium gallium галлий	Latin: *Gallia* = France
32	Ge	72.61	germanium Germanium germanium германий	Latin: *Germania* = Germany
33	As	74.922	arsenic	Greek: *arsenikon* = yellow orpiment (from Iranian: *zarnik* = gold-colored)
			Arsen arsenic мышьяк	*мышь* = mouse (for mouse poison)
34	Se	78.96	selenium Selen sélénium селен	Greek: *Selene* = moon
35	Br	79.904	bromine Brom brome бром	Greek: *bromos* = stench
36	Kr	83.80	krypton Krypton krypton криптон	Greek: *kryptos* = hidden
37	Rb	85.468	rubidium Rubidium rubidium рубидий	Latin: *rubidus* = red
38	Sr	87.62	strontium Strontium strontium стронтий	lead mines in Scottish *Strontian*

Chemical Elements (continued)

Atomic number	Symbol	Relative atomic mass	Name English German French Russian	Etymology E G F R
39	Y	88.906	yttrium Yttrium yttrium иттрий	Swedish town *Ytterby*
40	Zr	91.224	zirconium Zirkonium zirconium цирконий	Persian: *zargūn* = gold-colored
41	Nb	92.906	niobium Niob niobium ниобий	Greek: *Niobe* (daughter of Tantalus)
42	Mo	95.94	molybdenum Molybden molybdène молибден	Greek: *molybdos* = lead
43	Tc	97.907*	technetium Technetium technétium технеций	Greek: *technetos* = artificial
44	Ru	101.07	ruthenium Ruthenium ruthénium рутений	Medieval Latin: *Ruthenia* = Russia
45	Rh	102.906	rhodium Rhodium rhodium родий	Greek: *rhodon* = rose
46	Pd	106.42	palladium Palladium palladium палладий	asteroid *Pallas*

(continued)

Chemical Elements (continued)

Atomic number	Symbol	Relative atomic mass	Name *English* *German* *French* *Russian*	Etymology E G F R
47	Ag	107.868	silver Silber argent серебро	Gothic: *silubr* = silver like E Latin: *argentum* = silver Greek: *argyros* = silver Anatolian: *subauro* = shiny
48	Cd	112.411	cadmium Kadmium cadmium кадмий	Greek: *kadmeia* = calamine (zinc ore)
49	In	114.818	indium Indium indium индий	Greek: *indikon* = indigo
50	Sn	118.710	tin Zinn étain олово	Germanic: *tin;* possibly like F like E Latin: *stannum* = tin Indo-European root: *albhe* = white

51	Sb	121.757	antimony
			
			Antimon
			antimoine
			сурьма
				Arabic: *al ithmid*, Greek: *stimi,*
				  Coptic: *stem* (from this also
				  the Latin name *stibium*)
				
				Tatar: *sørmä*

52	Te	127.60	tellurium Tellur tellure теллур	Latin: *tellus* = earth
53	I	126.904	iodine Jod iode иод	Greek: *iodos* = violet
54	Xe	131.29	xenon Xenon xénon ксенон	Greek: *xenos* = foreign

Chemical Elements (continued)

Atomic number	Symbol	Relative atomic mass	Name *English* *German* *French* *Russian*	Etymology E G F R
55	Cs	132.905	cesium Zäsium césium цезий	Latin: *caesius* = bluish-gray
56	Ba	137.327	barium Barium barium барий	Greek: *barys* = heavy
57	La	138.906	lanthanum Lanthan lanthan лантан	Greek: *lanthanein* = to be hidden
58	Ce	140.115	cerium Zer cérium церии	asteroid *Ceres*
59	Pr	140.908	praseodymium Praseodym praséodyme празеодим	Greek: *prasios* = leek-green + *didymos* = twin
60	Nd	144.24	neodymium Neodym néodyme неодим	Greek: *neos* = new + *didymos* = twin
61	Pm	140.91*	promethium Promethium promethium прометий	Greek Titan *Prometheus*
62	Sm	150.36	samarium Samarium samarium самарий	Russian mining official V.Y. Samarski

(continued)

Chemical Elements (continued)

Atomic number	Symbol	Relative atomic mass	Name *English* *German* *French* *Russian*	Etymology E G F R
63	Eu	151.965	europium Europium europium европий	Latin: *Europa*
64	Gd	157.965	gadolinium Gadolinium gadolinium мадолиний	Finnish chemist J. *Gadolin*
65	Tb	158.925	terbium Terbium terbium тербий	Swedish town *Ytterby*
66	Dy	162.50	dysprosium Dysprosium dysprosium диспрозий	Greek: *dysprositos* = inaccessible
67	Ho	164.930	holmium Holmium holmium гольмий	Medieval Latin: *Holmia* = Stockholm
68	Er	167.26	erbium Erbium erbium эрбий	Swedish town *Ytterby*
69	Tu	168.934	thulium Thulium thulium тулий	Latin: *Thule* = Northland
70	Yb	173.04	ytterbium Ytterbium ytterbium иттербий	Swedish town *Ytterby*
71	Lu	174.967	lutetium Lutetium lutétium лютеций	*Lutetia* (Roman name for Paris)

Chemical Elements (continued)

Atomic number	Symbol	Relative atomic mass	Name *English* *German* *French* *Russian*	Etymology E G F R
72	Hf	178.49	hafnium Hafnium hafnium гафний	Medieval Latin: *Hafnia* = Copenhagen
73	Ta	180.948	tantalum Tantal tantale тантал	Greek: *Tantalos*, son of Zeus
74	W	183.84	tungsten Wolfram tungstène вольфрам	Swedish: = heavy stone German: = wolf's dirt like E like G
75	Re	186.207	rhenium Rhenium rhénium рений	Medieval Latin: *Rhenus* = Rhine
76	Os	190.23	osmium Osmium osmium осмий	Greek: *osme* = odor
77	Ir	192.22	iridium Iridium iridium иридий	Latin: *iris* = rainbow
78	Pt	197.08	platinum Platin platine платина	Spanish: *platina* = (small) silver
79	Au	196.966	gold Gold or золото	Indo-European root: *ġhel* = shining like E Latin: *aurum* = gold like E

(continued)

Chemical Elements (continued)

Atomic number	Symbol	Relative atomic mass	Name *English* *German* *French* *Russian*	Etymology E G F R
80	Hg	200.59	mercury	Roman god *Mercurius*
			Quecksilber	Old High German: *quecsilber* = living silver, mercury
			mercure	like E
			ртуть	Arabic: *utarid* = mercury
81	Tl	204.383	thallium	Greek: *thallos* = budding shoot
			Thallium	
			thallium	
			таллий	
82	Pb	207.2	lead	Indo-European root *ploud* = to flow
			Blei	Germanic: *bliwa*
			plomb	Latin: *plumbum*
			свинец	Church Slavic: *свиньчь*
83	Bi	208.980	bismuth	Old German: *Wiese* = meadow + *muoten* = to engage in wild mining
			Wismut	
			bismuth	
			висмут	
84	Po	208.980*	polonium	Modern Latin: *Polonia* = Poland
			Polonium	
			polonium	
			полоний	
85	At	209.987*	astatine	Greek: *astatos* = unstable
			Astat	
			astate	
			астатин	
86	Rn	222.018*	radon	*rad*(ium) emanation
			Radon	
			radon	
			эманация	
87	Fr	223.020*	francium	Modern Latin: *Francia* = France
			Francium	
			francium	
			франций	

Chemical Elements (continued)

Atomic number	Symbol	Relative atomic mass	Name *English* *German* *French* *Russian*	Etymology E G F R
88	Ra	226.025*	radium Radium radium радий	Latin: *radius* = ray
89	Ac	227.028*	actinium Aktinium actinium фктиний	Greek: *aktis* = ray
90	Th	232.0938*	thorium Thorium thorium торий	Old Norse: *Thor* = god of thunder
91	Pa	231.036*	protactinium Protaktinium protactinium протактиний	Greek: *protos* = first + *actinium*
92	U	238.051*	uranium Uran uranium уран	planet *Uranus*
93	Np	237.048*	neptunium Neptunium neptunium нептуний	planet *Neptune*
94	Pu	244.048*	plutonium Plutonium plutonium плутоний	planet *Pluto*
95	Am	243.061*	americium Americium américium америции	*America*
96	Cu	247.0704*	curium Curium curium кюрий	Pierre and Marie *Curie*

(continued)

Chemical Elements (continued)

Atomic number	Symbol	Relative atomic mass	Name *English* *German* *French* *Russian*	Etymology E G F R
97	Bk	247.070*	berkelium Berkelium berkelium беркелий	*Berkeley*, California
98	Cf	251.080*	californium Kalifornium californium калифорний	*California*
99	Es	252.083*	einsteinium Einsteinium einsteinium эйнштейний	Albert *Einstein*
100	Fm	257.095*	fermium Fermium fermium фермий	Enrico *Fermi*
101	Md	258.099*	mendelevium Mendelevium mendelevium менделевий	Dmitri Ivanovich *Mendeleyev*
102	No	259.01*	nobelium Nobelium nobelium нобелий	Alfred *Nobel*
103	Lr	260.105*	lawrencium Lawrencium lawrencium лоуренсий	Ernest *Lawrence*
104	Rf	261.109	rutherfordium Rutherfordium rutherfordium резерфордий	Ernest *Rutherford*
105	Db	262.114	dubnium Dubnium dubnium дубний	*Dubna*, town near Moscow

Chemical Elements (continued)

Atomic number	Symbol	Relative atomic mass	Name English German French Russian	Etymology E G F R
106	Sg	263.118	seaborgium Seaborgium seaborgium сиборгий	Glenn *Seaborg*
107	Bh	262.123	bohrium Bohrium bohrium борий	Niels *Bohr*
108	Hs	265.130	hassium	Latin: *Hassa* = Hessen, region around Darmstadt
			Hassium hassium хассий	
109	Mt	266.138	meitnerium Meitnerium meitnerium мейтнерий	Lise *Meitner*
110, 111, 112**				

*Only radioactive; the atomic mass of the most stable nuclide is given.
**No names have been approved yet for elements 110, 111, and 112.

Physical Chemistry

Name	Symbol	Units
heat	q, Q	J
work	w, W	J
internal energy	$U = q + w$	J
enthalpy	$H = U + pV$	J

[*Calorie* (cal) is still frequently used in spite of being an ill-defined unit. Usually 1 cal = 4.185 J.]

Name	Symbol	Units
thermodynamic temperature	T	K
Celsius temperature	$\theta, t = T - 273.15$	°C
entropy	$S \; (dS = dQ/T)$	J K^{-1}
Helmholtz energy	$F, A = U - TS$	J
Gibbs energy	$G = H - TS$	J
heat capacity		
at constant volume	$C_V = (\partial U/\partial T)_V$	J K^{-1}
at constant pressure	$C_p = (\partial H/\partial T)_p$	J K^{-1}
specific	$c_p = C_p/m$	J kg^{-1}K^{-1}
chemical potential (partial molar	$\mu_D = (\partial G/\partial n_D)_{T,p,nj \neq D}$	J mol^{-1}
Gibbs energy) of substance D		
standard chemical potential	$\mu^{\ominus}, \mu^{\circ}$	J mol^{-1}
affinity of reaction	$\mathcal{A}_D, A_D = -\Sigma \nu_D \mu_D = RT \ln K_{p,T}$	J mol^{-1}
equilibrium constant	$K^{\ominus}, K = \exp(-\Delta G^{\circ}/RT)$	1
acid dissociation constant		
thermodynamic	K_a	mol L^{-1}
apparent	K'_a	mol L^{-1}
basic dissociation constant		
thermodynamic	K_b	mol L^{-1}
apparent	K'_b	mol L^{-1}
ion product of water	K_w	(mol L^{-1})^2
activity coefficient	$f_D = a_D/x_D$	1
(according to Raoult's law)		

[This applies to pure phases and substances in mixtures or solvents.]

Name	Symbol	Units
activity coefficient	$\gamma_c = a_{c,D}c^{\ominus}/c_D$	1
(according to Henry's law)		
activation energy	E_a	J
ionic strength	$I_c, I = \frac{1}{2}\Sigma c_D z_D^2$	mol L^{-1}

Physical Chemistry *(continued)*

Name	Symbol	Units
charge number of an ion	z_D	1
osmotic pressure	$\pi = c_D RT$	Pa
rate of conversion	$d\xi /dt$	mol s^{-1}
rate of concentration change	$r_D, v_D = dc_D/dt$	mol L^{-1} s^{-1}
rate of reaction	$v = v_D^{-1} dc_D/dt$	mol L^{-1} s^{-1}
rate constant	k	(L mol$^{-1})^{n-1}$ s^{-1}
overall order of reaction	$n = \sum n_D$	1
half life	$t_{1/2}$	s
relaxation time	τ	s
collision diameter	$d_{ij} = r_i + r_j$	m
collision cross section	$\sigma = \pi d_{ij}^{2}$	m^2
collision frequency (of A with B)	$z_A(B) = C_B \bar{\sigma} \bar{c}_{A,B}$	s^{-1}
mean relative speed between A and B (μ is reduced mass)	$\bar{c}_{AB} = (8kT/\pi\mu)^{1/2}$	m s^{-1}
mean free path	$\lambda = \bar{c}/z_A$	m
quantum yield	$\emptyset$	1
[This is the ratio of the number of generated events to the number of photons absorbed.]		
photochemical yield	$\emptyset$	1
[This is generally the ratio of the rate of reaction to the rate of photon absorption.]		
electromotive force	$E, \text{emf} = \lim_{I=0} \Delta V$	V
standard emf	$E^{\ominus} = -\Delta G^{\ominus}/nF = (RT/nF) \ln K^{\ominus}$	V
oxidation-reduction (redox) potential	E_h	V
standard redox potential (half-reduction potential, midpoint potential)	$E^{\ominus}, E^{\circ}, E_m$	V
standard redox potential at specified pH	$E^{\ominus\prime}, E^{\circ\prime}, E_m'$	V
pH	$\approx \log_{10} [H^+]$	1
buffering capacity	$\beta = d[B]/d\,pH$	mol L^{-1}
charge number	n, v_e, z	1
surface electric potential	$\chi = \emptyset - \psi$	V
Galvani potential difference	$\Delta\emptyset = \emptyset^{\beta} - \emptyset^{\alpha}$	V
[This is the potential difference between points in bulk phases α and β of identical composition.]		
Volta potential difference	$\Delta\psi = \psi^{\beta} - \psi^{\alpha}$	V
[This is the potential difference caused by electric charge on phase α and β.]		
electrochemical potential (of substance D)	$\tilde{\mu} = (\partial G/\partial n_D)$	J mol^{-1}

Physical Chemistry *(continued)*

Name	Symbol	Units
protonmotive force	$\text{pmf} = n\Delta\psi - 2.3RT\Delta\text{pH}/F$	V
current density	$j = I/A$	A m^{-2}
surface charge density	$\sigma = Q/A$	C m^{-2}
thickness of diffusion layer (for substance D)	$\delta_D = D_D/k_{d,D}$	m
diffusion rate constant (of substance D)	$k_{d,D} = v_D I_{\lim,D}/nFc_D A$	m s^{-1}
overpotential (zeta potential)	$\eta, \zeta = E_I - E_{I=0} - IR$	V
conductivity	$\kappa = j/E$	S m^{-1}
molar		
of substance D	$\Lambda_D = \kappa/c_D$	S m^2 mol^{-1}
of ion C	$\lambda_C = z_C F\mu_C$	S m^2 mol^{-1}
electric mobility (of ion C)	$\mu_C = v_C/E$	m^{2}V^{-1} s^{-1}
reflection coefficient	σ	1
dissipation function	Φ	J s^{-1}
area per molecule (of substance D)	$a_D = A/N_D$	m^2
film thickness	t, h, δ	m
flux (of quantity X)	$J_X = A^{-1}dX/dt$	(X) m^{-2} s^{-1}
volume flow rate	$q_V = dV/dt$	m^3 s^{-1}
mass flow rate	$q_m = dm/dt$	kg s^{-1}
diffusion coefficient	D	m^2 s^{-1}
permeability coefficient	P	m s^{-1}
partition coefficient	K	1
critical coagulation concentration	ccc	mol L^{-1}
critical micellization concentration	cmc	mol L^{-1}
isoelectric point	iep, pI	(pH)
order parameter (of membranes)	S	1
transition temperature (of membranes)	T_t	K
bilayer thickness	d_l	m
radius of curvature	R	m

Organic and Structural Chemistry

Numbering of carbon atoms in molecules should follow these rules:

1. The number of C atoms in a molecule is designated with a subscript, thus C_{14} for a molecule containing 14 C atoms.
2. The position of the C atom (usually in connection with a substituent on it) is designated with a subscript in parentheses, thus $C_{(14)}$ for the 14th carbon atom in a chain. Still, more frequently one finds designations of C atom positions using a hyphen, thus C-3, and the like.

To describe **torsion angles** about two adjacent (usually C) atoms the following abbreviations are used: if the angle is equal to 0° it is *cis*, or c̓; if the angle is equal to 60°, it is *gauche* or *skew*, or g; if it is equal to 180°, it is *trans*, or t.

To describe absolute molecular chirality (or handedness) of an atom the following rule is observed. If the chiral atom is viewed from the side opposite to the substituent with the lowest atomic number and the path traced from the substituent with the highest atomic number to the one with the next highest atomic number goes clockwise, the chiral number is (*Re*), if it goes counterclockwise it is (*Si*). Similarly, *E* and *Z* are used for designation of double bonds.

Some of the **structural prefixes** are written in italics, such as *cis-*, *trans-*, *dextro-*, *levo-*, *meso-*, *o-* (for *ortho-*), *m-* (for *meta-*), *p-* (for *para-*), *rac-* (for racemic mixture), *syn-*, *anti-*, *endo-*, *exo-*, *inter-*, *ent-*, *ambo-*, *all-*, *nido-*, *abeo-*, *catena-*; likewise, locants referring to various atoms are written in italics, thus *N-*, *O-*, *S-*. Roman (upright) letters are used in the following cases: allo-, bis-, tris-, c- for c̲yclo- (however, in abbreviations: *cyclo*-C_{30}), i- for i̲so-, n- for n̲ormal, s- for s̲econdary, t- (for t̲ertiary), epi-, neo-, nor-, spiro-, seco-, homo-, and des-. All designations of position using Greek letters should be in roman type (e.g., α, β, ω). Designations of fused-ring compounds should be printed in italic letters within brackets, such as [*a*], [*de*], [*j*], [*rst*].

Isotopic substitution is shown either with the nuclide designation in brackets, such as *S*-[8-^{14}C]adenosyl[^{35}S]methionine, or, wherever possible, more simply, e.g. $H_2{}^{35}SO_4$. In this context, letter G stands for general labeling, letter U stands for universal labeling, e.g. [G-^{3}H]octanoic acid, or [U-^{14}C]hexanoic acid.

The symbol @ has been used to indicate the presence of a noncarbon atom in the molecule of fullerenes, e.g., La@C$_{82}${C$_{3V}$(II)}; however, it is now recommended to use the prefix *incar* instead, abbreviated *i*, thus *i*LaC$_{82}${C$_{3V}$(II)}.

Reaction mechanisms are designated in a special way; thus, S$_N$1 for a unimolecular, nucleophilic substitution reaction, and S$_N$2 for a bimolecular, nucleophilic substitution reaction.

In **coordination chemistry**, three Greek letters are in general use: η (read *hapto*) is a prefix to the ligand name indicating connectivity between the ligand and the central atom, its right numerical superscript indicating the number of coordinating atoms in the ligand that bind to the metal; κ (read *kappa*), as a prefix to an italic symbol of the element, designates a simple atom ligation to a coordination center; μ (read *mu*) stands for a bridging ligand between two or more central atoms.

For copolymers, i.e., polymers based on more than one kind of monomer, the following designations are used:

-alt-	alternating	*-per-*	periodic
-block-	block	*-ran-*	random
-co-	unspecified	*-stat-*	statistical
-graft-	graft		

Many **substituents** (in fact, radicals, designated generically with R) in organic chemistry are routinely abbreviated and many of them (those marked with an asterisk) need not be explained.

*Ac-	acetyl	Lin-	linoleoyl
Ach-	arachidoyl	αLnn	linolenoyl
Δ$_4$Ach-	arachidonoyl	Mal<	maleoyl
Beh-, Behen-	behenoyl	Mal-	maloyl
Br-	butyryl	*Me-	methyl
*Bu-	butyl	Me$_3$Si	trimethylsilyl
Bz-	benzoyl	MeO-	methoxy
Bzl-	benzyl	Mon-	montanoyl
Cbm-	carbamoyl	Myr-	myristoyl
Cet	carboxyethyl	Ner-	nervonoyl
Cm-	carboxymethyl	Nn-	nonyl
Crt-	cerotoyl	Oc-	octyl
*d-	deoxy	Oco-	octanoyl

Dec-	decyl	Ole-	oleoyl
Dnp-	dinitrophenyl	*P-	phosphoryl
Dns-	dansyl	Pam-	palmitoyl
Dod-	dodecyl	Pe-	pentyl
*Et-	ethyl	*Ph-	phenyl
Eto-	ethoxy	-Ph₃C-	triphenylmethyl
For- (or HCO)	formyl	Ph<	phthaloyl
Gc-	glycoloyl	Pp-	propionyl
GlcA-	gluconyl	Pr-	propyl
GlcU-	glucuronyl	Ste-	stearoyl
H$_2$-	dihydro	Suc-	succinyl
H$_4$-	tetrahydro	*Tos-	tosyl
Hp-	heptyl	Trt-	trityl
Hpo-	heptanoyl	Und-	undecyl
Hx-	hexyl	Vac-	vaccenoyl
cHx-	cyclohexyl	Vl-	valeryl
Lau-	lauroyl	iVl-	isovaleryl
Lignocer-	lignoceroyl		

Biochemistry and Molecular Biology

Various groups of compounds of importance in biochemistry have their names abbreviated, some of them codified to the extent that the abbreviations need no explanation (these are marked with an asterisk).

Amino Acids

*alanine	Ala	A	*glutamine	Gln	Q
*β-alanine	βAla		*Glu *or* Gln	Glx	Z
alloleucine	Alle		*glycine	Gly	G
2-aminoadipic acid	Aad		*histidine	His	H
3-aminoadipic acid	βAad		homocysteine	Hcy	
2-aminobutyric acid	Abu		homoserine	Hse	
4-aminobutyric acid	γAbu,GABA		hydroxylysine	Hyl	
(γ-amino)butyric acid			hydroxyproline	Hyp	
2-aminohexanoic acid	Ahx		*isoleucine	Ile	I
6-aminohexanoic acid	εAhx		*leucine	Leu	L
2-aminoisobutyric acid	Aib		*lysine	Lys	K
2-aminopimelic acid	Apm		methionine	Met	M
2-aminovaleric acid	Ave, Ape		norleucine	Nle	
δ-aminolevulinic acid	DALA		norvaline	Nva	
*arginine	Arg	R	ornithine	Orn	
*asparagine	Asn	N	*phenylalanine	Phe	F
*aspartic acid	Asp	D	phosphoserine	*P*-Ser	
*Asn *or* Asp	Asx	B	*proline	Pro	P
4-carboxyglutamic acid	Gla		pyroglutamic acid	Glp	
citrulline	Cit		sarcosine	Sar	
cysteic acid	Cya		selenocysteine[a]	Sec	U
*cysteine	Cys	C	*serine	Ser	S
2,4-diaminobenzoic acid	DABA		*threonine	Thr	T
2,4-diaminobutyric acid	Dab, DABA		thyroxine	Thx	
2,6-diaminopimelic acid	Dpm, DAP		*tryptophan	Trp	W
dihydroxyphenylalanine	DOPA		*tyrosine	Tyr	Y
*glutamic acid	Glu	E	*valine	Val	V

[a]Recommended by the IUBMB-IUPAC
Joint Commission on Biochemical Nomenclature

The sequence number of an amino acid in a polypeptide chain was first written as a right superscript (Asn[125]), then most often as a number following the abbreviation, either without a space (Asn125) or with a space (Asn 125). Now it appears most acceptable, for reasons of both analogy with similar abbreviations and of typography, to use a hyphen (Asn-125).

Sugars and Derivatives

abequose	Abe	iduronic acid	IdoA
allose	All	inositol	I
altrose	Alt	*myo*-inositol	Ins
*arabinose	Ara	*lactose	Lac
2,3-bis(phospho)-D-glycerate	BPG	lyxose	Lyx
*fructose	Fru	*maltose	Mal
*fucose	Fuc	*mannose	Man
*galactose	Gal	muramic acid	Mur
*gluconic acid	GlcA	neuraminic acid	Neu
*glucosamine	GlcN	*N-acetylneuraminic acid	NANA
*N-acetylglucosamine	GlcNAc	quinovose	Qui
*glucose	Glc	*rhamnose	Rha
*2-deoxyglucose	dGlc	2-deoxyribose	dRib
*glucuronic acid	GlcU	*ribose	Rib
glyceraldehyde	Gra	ribulose	Rbu, Rul
glyceric acid	Gri	sedoheptulose	Sed
glycerol	Gro	*sucrose	Suc, Sac
glycerone (dihydroxyacetone)	Grn	talose	Tal
gulose	Gul	*trehalose	Tre
idose	Ido	*xylose	Xyl
		xylulose	Xlu, Xul

In sugar stereochemistry the following abbreviations are used to designate structure: *allo-, altro-, arabino-, chiro-, epi-, erythro-, galacto-, gluco-, glycero-, gulo-, ido-, lyxo-, manno-, muco-, myo-, neo-, ribo-, scyllo-, talo-, threo-,* and *xylo-. B* stands for boat configuration, *C* for chair configuration, H_5 for half-chair configuration, *p* for pyranose, *f* for furanose, *sn* stands for stereospecific numbering.

To describe the **position of the hydroxyl group** at the highest-numbered carbon atom of monosaccharides, the following rule is observed. If the hydroxyl group is to the right of the vertical in a Fischer projection, the sugar will be D; if it is to the left, the sugar will be L.

In complex sugars and glycoproteins, the saccharide moieties are linked to either an oxygen or to a nitrogen atom of the amino acid residue (serine, threonine, and hydroxylysine to O, and asparagine to N) of the peptide chain. Such sugars are termed O- or N-linked. If, however, these or any other symbols for atoms are used as locants, they should be italicized, e.g., *N*-glycolyl, *O*-acetyl, *S*-methyl, even if this practice appears to unnecessarily complicate matters for printers.

Special mention should be made of phosphorylated sugars, such as occur in a number of metabolic pathways. A more rigorous usage recommends the three-letter code, such as Rib-5-*P* for D-ribose 5-phosphate, or Fru-1,6-*P*$_2$ for D-fructose 1,6-bisphosphate. However, even more contracted abbreviations do occur, particularly in graphic representations of the reaction schemes, such as R5P and F1,6P$_2$ (or even FBP) for the above two examples; similarly, S7P for sedoheptulose 7-phosphate, X5P for xylulose 5-phosphate, or E4P for erythrose 4-phosphate. For obvious reasons, these abbreviations are not to be used without proper explanation.

Nucleic Acids, Their Components and Derivatives

A, Ade	adenine
A, Ado	adenosine
ADP[S]	adenosine 5'-β-thiodiphosphate
AIR	5-aminoimidazole ribonucleotide
AMP-PNP	5'-adenylyl-β,γ-imidophosphate
AMP, ADP, ATP	adenosine mono-, di-, and triphosphate

[Analogously, CMP, CDP, CTP for cytidine phosphates, GMP, GDP, GTP for guanosine phosphates. IMP, IDP, ITP for inosine phosphates, OMP, ODP, OTP for orotidine phosphates, UMP, UDP, UTP for uridine phosphates, and XMP, XDP, XTP for xanthosine phosphates.]

cAMP	cyclic AMP (analogously cGMP, cIMP)
APS	adenosine 5'-phosphosulfate
AZT	3'-azido-3'-deoxythymidine
B	cytosine *or* guanine *or* thymine
C, Cyd	cytidine
C, Cyt	cytosine
D	adenine *or* guanine *or* thymine
dA, dC, dG, dT, dU (dAdo, dCyd, dGuo, dThd, dUrd)	deoxyadenosine, deoxycytidine, deoxyguanosine, thymidine (!), deoxyuridine

[dCDP, dCTP, etc. are used for the deoxy derivatives of nucleoside phosphates.]

ddTTP	2',3'-dideoxythymidine triphosphate
DHU	dihydrouridine
DNA	deoxyribonucleic acid
amDNA	anti-messenger DNA

(continued)

Nucleic Acids, Their Components and Derivatives (continued)

cDNA	complementary DNA
cccDNA	covalently closed circular DNA
ctDNA	chloroplast DNA
dsDNA	double-stranded DNA
hDNA	hybrid DNA
iDNA	intercalary DNA
kDNA	kinetoplast DNA
msDNA	multicopy single-stranded DNA
mtDNA	mitochondrial DNA
nDNA	nuclear DNA
pDNA	plasmid DNA
rDNA	ribosomal DNA
recDNA	recombinant DNA
rfDNA	replicative-form DNA
scDNA	single-copy DNA
ssDNA	single-stranded DNA
tDNA	transfer DNA
T-DNA	transfer DNA of Ti plasmid

[For different forms of DNA, upright prefixes A-, B-, C- and Z- are used.]

G, Gua	guanine
G, Guo	guanosine
H	adenine *or* cytosine *or* thymine
Hyp	hypoxanthine
I, Ino	inosine
K	guanine *or* thymine
M	adenine *or* cytosine
N	adenine *or* cytosine *or* guanine *or* thymine
O, Ord	orotidine
O, Oro	orotate
PAP	3′-phosphoadenosine-5′-phosphate
PAPS	3′-phosphoadenosine-5′-phosphosulfate
poly(A)	simple homopolymer of A
poly(dA-dT)	alternating co-polymer of dA and dT
Pu	purine
Py	pyrimidine
Q	queuosine (7-[4,5-*cis*-dihydroxy-1-cyclopentene-3-aminomethyl]-7-deazaguanosine)
R	adenine *or* guanine
RNA	ribonucleic acid

Nucleic Acids, Their Components and Derivatives (continued)

cRNA	complementary RNA, copy RNA
ctRNA	chloroplast RNA; countertranscript RNA
dsRNA	double-stranded RNA
gRNA	(trans-acting) guide RNA
hnRNA	heterogeneous nuclear RNA
mRNA	messenger RNA
micRNA	messenger-RNA-interfering complementary RNA
mtRNA	mitochondrial RNA
nRNA	nuclear RNA
rRNA	ribosomal RNA
scRNA	small cytoplasmic RNA
snRNA	small nuclear RNA
snoRNA	small nucleolar RNA
ssRNA	single-stranded RNA
tRNA	transfer RNA
tcRNA	translational-control RNA

[For different forms of RNA, upright prefixes A- and Z- are used.]

RNP	ribonucleoprotein
S	guanine *or* cytosine
Sno	thioinosine
SU	thiouridine
Syp	mercaptopurine
T, Thd	ribosylthymine (!)
Thy	thymine
U, Ura	uracil
U, Urd	uridine
W	adenine *or* thymine
X, Xan	xanthine
X, Xao	xanthosine
Y	cytosine *or* thymine
Ψ, Ψrd	pseudouridine (= 5-ribosyluracil)

For substituents in bases the following letters placed directly before the single-letter symbol (such as A, C, G, U) are to be used:

m, e, ac	methyl, ethyl, acetyl	c	deaza (N→C)	f	formyl
n	amino (H→N)	h	dihydro	i	isopentenyl
d	deamino (N→O)	oh	hydroxy	s	thio, mercapto
z aza	(C→N)	aa	aminoacyl		

Plant Lectins

Abbreviation	Source
ConA (concanavalin A) from	*Canavalia ensiformis* (jack bean)
DBA	*Dolichos lablab* (hyacinth bean)
LCA	*Lens culinaris* (lentil)
PHA	*Phaseolus vulgaris* (kidney bean)
PNA	*Arachis hypogaea* (peanut)
PSA	*Pisum sativum* (garden pea)
RCA	*Ricinus communis* (castor bean)
SBA	*Glycine max* (soybean)
SJA	*Sophora japonica* (Japanese pagoda tree)
UEA	*Ulex europaeus* (common gorse)
WGA	*Triticum vulgare* (wheat)

Vitamins

A	retinol, retinal, retinoic acid
B_1	thiamin
B_2	riboflavin
B_6	pyridoxal, pyridoxamine, pyridoxine
B_{12}	cobalamin
C	ascorbic acid
D	calcidiol or $1\alpha,25$-dihydroxycholecalciferol (for vitamin D_2)
E	α-tocopherol
K	phylloquinone (for vitamin K_1)

Note that for vitamins of the B_{12} family hyphenated forms are now preferred, to make further distinction easier, e.g., vitamin B-12r, vitamin B-12s, etc. With other vitamins, a subscript is used for their subtypes, e.g., vitamin D_2, vitamin K_1, and the like.

Some of the symbols are now rarely used, such as B_3 for pantothenic acid, B_5 for nicotinamide, B_7 for carnitine, and H for histidine. In the past, symbols B_4, B_8, B_9, B_{10}, B_{11}, B_{13}, B_{15}, B_{17}, B_a, B_p, B_t, B_w, B_x, F, G, I, J, L, M, P, PP, R, S, T, U, and V were occasionally introduced for vitamin-like activities, but the corresponding substances were later found to be either mixtures of vitamins or compounds of non-vitamin nature.

Miscellaneous Abbreviations in Biochemistry

A23187	a calcium ionophore
ABA	abscisic acid
ABC (transporters)	ATP-binding-cassette proteins
ACh	acetylcholine
ACP	acyl carrier protein
AFP	α-fetoprotein

(continued)

Miscellaneous Abbreviations in Biochemistry (continued)

Apo	apoliprotein
ARF, Arf	ADP-ribosylation factor
Bchl	bacteriochlorophyll
BoNT	botulinum neurotoxin
Bph	bacteriopheophytin
C_3 cycle	reductive pentose phosphate cycle
CaM	calmodulin
LCAM, NCAM	molecule involved in cell adhesion (for liver and neuron, respectively)
CAP	catabolite activator protein
CCK	cholecystokinin
Cer	ceramide
Ch, Cho	choline
CHIP	channel-like integral membrane protein
Chl	chlorophyll
CM	chloramphenicol
CoA	coenzyme A
CoM	coenzyme M
$(\alpha, \beta, \gamma, \delta)$ COP	nonclathrin vesicle coat proteins (coatamers)
cpn	chaperonin
Cr	creatine, creatinine
CRP	cAMP-receptor protein
CsA	cyclosporin A
CT	calcitonin
cyt	cytochrome
DAG	diacylglycerol
DES	diethylstilbestrol
dgl	desmosome glycoprotein
DIC	differential interference contrast (Nomarski microscopy)
Dnp	dinitrophenyl
DNP	dinitrophenol
Dol	dolichol
dpl	desmosome protein
DPG	diphosphatidylglycerol (cardiolipin)
EC	adenylate energy charge; Enzyme Commission
EF(-hand)	a calcium-binding motif
EPS	extracellular polysaccharide
ETC	electron-transport chain
Etn	ethanolamine

(continued)

Miscellaneous Abbreviations in Biochemistry (continued)

ET	endothelin
F_{420}, F_{430}	factors 420 and 430 (in methanogenesis)
FABP	fatty-acid-binding protein
FAD	flavin-adenine dinucleotide
Fd	ferredoxin
FH_4	tetrahydrofolic acid
FMN	flavin mononucleotide (= riboflavin 5'-phosphate)
FN	fibronectin
F_oF_1-ATPase	F-type proton ATPase, ATP synthase
G-proteins	GTP-binding proteins
GAP	GTPase-activating protein
GEF	guanine-nucleotide exchange factor
GIP	GTPase-inhibiting protein; general insertion protein
GIPL	glycoinositolphospholipid
GRF	guanine-nucleotide release factor
GSH, GSSG	reduced and oxidized glutathione
H1, H2A. H2B, etc.	classes of histones
H_I, H_{II}	hexagonal I and hexagonal II lipid phases
Hb	hemoglobin
HbA	adult Hb
HbF	fetal Hb
HbS	sickle-cell Hb
HDL	high-density lipoprotein
HMM	heavy meromyosin
HSA	human serum albumin
hsp	heat-shock protein
IDL	intermediate-density lipoprotein
INH	isoniazid
JAK	Janus kinase (*or* just another kinase)
K	phylloquinone
L	lamellar lipid phase
LAM	lipoarabinomannan
LAR	ligase amplification reaction
LDL	low-density lipoprotein
LHC	light-harvesting complex
LMM	light meromyosin
$Lip(S_2)$	α-lipoic acid
LPA	lysophosphatidic acid
LPC	lysophosphatidylcholine
LPS	lipopolysaccharide

Miscellaneous Abbreviations in Biochemistry (continued)

LSD	lysergic acid diethylamide
LT	leukotriene
LV	lipovitellin
LX	lipoxin
MAG	myelin-associated glycoprotein
MAP	microtubule-associated protein; mitogen-activated protein
MARP	microtubule-associated repetitive protein
Mb	myoglobin
MCP	methyl-accepting chemotaxis protein
MetHb	methemoglobin
MHC	myosin heavy chain
MIP	major intrinsic protein (e.g., of bovine lens)
MK	menaquinone
MLC	myosin light chain
NAD	nicotinamide-adenine dinucleotide
NADP	nicotinamide-adenine dinucleotide phosphate
NMN	nicotinamine mononucleotide
NSF	N-ethylmaleimide-sensitive fusion (protein)
Omp	outer membrane protein
OSCP	oligomycin-sensitivity-conferring protein
P	rippled lipid phase
P680, P700	pigments absorbing light at 680 and 700 nm
PA	phosphatidic acid
PAS	p-aminosalicylic acid
PBG	porphobilinogen
PC	phosphatidylcholine (lecithin)
PE	phosphatidylethanolamine
PEP	phosphoenolpyruvate
PG	phosphatidylglycerol; prostaglandin; peptidoglycan
PGA_2, PGB_1	prostaglandins A_2 and B_1
PG-1, PG-2, etc.	protegrins
PHA	phytohemagglutinin
PHB	poly-β-hydroxybutyrate
PI	phosphatidylinositol
PL	pyridoxal; phospholipid
PM	pyridoxamine
PN	pyridoxine
PrP^C, PrP^{Sc}	prion proteins
PS	phosphatidylserine

(continued)

Miscellaneous Abbreviations in Biochemistry (continued)

Ptd	phosphatidyl group
Pte	pteroic acid
PTS	phosphotransferase system
Pxy	pyridoxyl
Q	cubic lipid phase; ubiquinone
Q_{10}	ratio of reaction rate at T to that at $(T-10)$
RTK	receptor tyrosine kinase
SAM	S-adenosylmethionine
SNAP	soluble NSF-attachment protein
SNARE	SNAP receptor
T, Tα, T$\beta\gamma$	transducin and its subunits
TAG	triacylglycerol
TEA	triethanolamine
TeTx	tetanus toxin
THMP	tetrahydromethanopterin
TIP	tonoplast intrinsic protein
TRAMP	tyrosine-rich acidic matrix protein
TTX	tetrodotoxin
Tx, TX	thromboxane
UQ	ubiquinone
VAMP	vesicle-associated membrane protein
VHDL	very-high-density lipoprotein

Miscellaneous Abbreviations in Molecular Biology

A arm	anticodon-containing arm of tRNA
A site	aminoacyl-recognition or acceptor site on a ribosome
ACE	amplification control element
ANK (repeat)	amino acid motif found in ankyrins
ARS	autonomously replicating sequence
ATD	amino-terminal domain
bp	base pair
$\bar{c}$ (c value)	amount of DNA in a haploid genome (in pg or M_r)
C, G, P segments	parts of DNA that are inverted during site-specific recombination by a specific "recombinase"
$c_0 t$	product of concentration of denatured DNA prior to association and of the time that reassociation has taken place [occasionally used as cot when referring to the technique]
CCA arm	amino acid acceptor arm of tRNA
cM	centimorgan
CTD	carboxy-terminal domain

(continued)

Miscellaneous Abbreviations in Molecular Biology (continued)

D arm	dihydrouridine-containing arm of tRNA
D loop	a loop of dsDNA formed by displacement of a strand segment by a homologous segment of DNA or RNA complementing it
DP	degree of polymerization
DR	direct repeat
E site	exit site on a ribosome
EF-Ts, EF-Tu	elongation factors of bacteria
EST	expressed sequence tag
GC%	ratio of (G + C) to (A + T) in DNA
gp	gene product
gp56	glycoprotein (of 56 kDa molar mass)
HLX	helix-turn-helix
I elements	transposons of *Drosophila* causing dysgenesis of male- and female-type hybrids
IF-1, IF-2	initiation factors (of bacteria)
eIF-4B, eIF-3β	initiation factors (of eukaryotic cells)
IncC, IncF, etc.	incompatibility groups in plasmids
IR	inverted repeat
IS*3*, IS*12*	insertion sequence
IVS	intervening sequence
kb	kilobase
kbp	kilobase pair
L, Lk	linking number
LTR	long terminal repeat
ORF	open reading frame
p35	protein (of 35 kDa molar mass)
pp40	phosphoprotein (of 40 kDa molar mass)
P elements	transposons of *Drosophila* causing dysgenesis of hybrids
P site	peptidyl or donor site on a ribosome
PEST	sequence of Pro-Glu-Ser-Thr involved in protein targeting
PU	palindromic unit
r (as prefix)	recombinant
R loop	loop between RNA and DNA strand at start of transcription
RAPD	random amplification of polymorphic DNA
RE	restriction endonuclease
RF	termination factor
SD	Shine-Dalgarno sequence
SINE	short interspersed element of DNA
SL	selectivity factor

(continued)

Miscellaneous Abbreviations in Molecular Biology (continued)

SRP	signal-recognition particle
T arm	pseudouridine-containing arm of tRNA
T, Tw	twist
TAF	TBP-associated factor
TBP	TATA-binding protein
TE	transposable element
TF	transcription factor
T_m	melting temperature
Tn9, Tn102	eukaryotic transposons
Tn*3*, Tn*2410*	bacterial transposons
UAS	upstream activating sequence
UBF	upstream binding factor
UCE	upstream control element
URS	upstream regulating sequence
UTR	untranslated region
W, Wr	writhing number
ZIP	leucine zipper domain
α	winding number of duplex DNA
β	number of secondary turns in duplex DNA

Methods and Techniques In Biophysics and Biochemistry

AAS	atomic absorption spectroscopy
AES	Auger electron spectroscopy
ARPES	angular resolved photoelectron spectroscopy
ATLC	adsorption thin-layer chromatography
BLM	black lipid membrane
CACE	counteracting chromatographic electrophoresis
CARS	coherent anti-Stokes Raman spectroscopy
CAT	computerized axial tomography
CD	circular dichroism
CEP	counterelectrophoresis
CFZC	continuous-flow zonal centrifugation
CIDEP	chemically induced dynamic electron polarization
CIDNP	chemically induced dynamic nuclear polarization
CIE	countercurrent immunoelectrophoresis
CIMS	chemical ionization mass spectrometry
CLEC	chiral ligand exchange chromatography
COSY	correlation spectroscopy

(continued)

Methods and Techniques In Biophysics and Biochemistry (continued)

CPC	centrifugal partition chromatography
CSLM	confocal scanning light (*or* laser) microscopy
DICD	dispersion-induced circular dichroism
DIM	digital imaging microscopy
DMDC	dimethyldithiocarbamate
DRESS	depth-resolved surface coil spectroscopy
DRIFT	diffuse reflectance infrared Fourier transform
DSC	differential scanning calorimetry
DTA	differential thermal analysis
EAPFS	electron appearance potential fine structure
EC	exclusion chromatography
EDAX	energy dispersive analysis by X-rays
EELS	electron-energy-loss spectroscopy
EIA	enzyme immunoassay
EID	electroimmunodiffusion
EIMS	electron ionization mass spectrometry
ELISA	enzyme-linked immunosorbent assay
EM	electron microscopy
EMER	electromagnetic molecular electronic resonance
EMT	enzyme-multiplied immunoassay technique
ENDOR	electron-nuclear double resonance
EPMA	electron-probe X-ray microanalysis
EPR	electron paramagnetic resonance
ESCA	electron spectroscopy for chemical analysis
ESE	electron spin-echo (spectroscopy)
ESR	electron spin resonance
ETLC	extraction thin-layer chromatography
ETS	electron tunneling spectroscopy
EXAFS	extended X-ray absorption fine structure (spectroscopy)
EXELFS	extended electron-loss fine structure (spectroscopy)
FACS	fluorescence-activated cell sorter
FEISEM	field emission in-lens scanning electron microscopy
FIA	fluorescence immunoassay
FIM	field-ion microscopy
FIR	far infrared (spectroscopy)
FM	fluorescence microscopy
FPLC	fast protein liquid chromatography
FRAP	fluorescence recovery after photobleaching

(continued)

Methods and Techniques In Biophysics and Biochemistry (continued)

FRAT	free radical assay technique
FRET	Förster resonance energy transfer
FTIR	Fourier-transform infrared (spectroscopy)
FWHH	full width at half height (spectrometry)
GC	gas chromatography
GFC	gel filtration chromatography
GGE	gradient gel electrophoresis
GLC	gas-liquid chromatography
GLPC	gas-liquid partition chromatography
GPC	gas-permeation chromatography
HPLC	high-performance liquid chromatography
HPSEC	high-performance size-exclusion chromatography
HPTLC	high-performace thin-layer chromatography
HRC	high-resolution chromatography
HTRS	high-temperature reflectance spectroscopy
HVEM	high-voltage electron microscopy
HVPE	high-voltage paper electrophoresis
HVTEM	high-voltage transmission electron microscopy
ICD	induced circular dichroism
IEC	ion-exchange chromatography
IEF	isoelectric focusing
IEM	immunoelectron microscopy
IEOP	immunoelectroosmophoresis
IEP	immunoelectrophoresis
IETS	inelastic electron tunneling spectroscopy
IPC	ion-pair chromatography
IR	infrared (spectroscopy)
IRMA	immunoradiometric assay
ITLC	instant thin-layer chromatography
LC	liquid chromatography
LEC	ligand exchange chromatography
LEED	low-energy electron diffraction
LEIS	low-energy ion scattering
LLC	liquid-liquid chromatography
LMR	laser magnetic resonance
LSC	liquid-solid chromatography
MAIA	magnetic immunoassay
MC	Monte Carlo (method)
MCD	magnetic circular dichroism

Methods and Techniques In Biophysics and Biochemistry (continued)

MD	molecular dynamics
MLEE	multi-locus enzyme electrophoresis
MM	molecular mechanics
MS	mass spectrometry
MW	microwave (spectroscopy)
NIR	near infrared (spectroscopy)
NMR	nuclear magnetic resonance
NOE	nuclear Overhauser effect (*or* enhancement)
NOESY	nuclear Overhauser effect spectroscopy
NQR	nuclear quadrupole resonance
ODMR	optically detected magnetic resonance
OFAGE	orthogonal-field-alternation gel electrophoresis
ORD	optical rotatory dispersion
PAGE	polyacrylamide gel electrophoresis
PARAP	polarized absorption recovery after photobleaching
PC	paper chromatography
PCR	polymerase chain reaction
PET	positron emission tomography
PFGE	pulsed-field gel electrophoresis
PIXE	proton-induced X-ray emission
PRFT	partially relaxed Fourier transform
PS	photoelectron spectroscopy
PTLC	preparative thin-layer chromatography; precipitation thin-layer chromatography
RAST	radioallergosorbent test
REM	reflection electron microscopy
RGC	radio-gas chromatography; reaction gas chromatography
RHEED	reflection high-energy electron diffraction
RIA	radio-immunoassay
RIST	radio-immunosorbent test
RNAA	radiochemical neutron activation analysis
ROE	rotating-frame Overhauser enhancement
RPC	reversed-phase chromatography
RPHLPC	reversed-phase high-performance liquid chromatography
RPLC	reversed-phase liquid chromatography
RRS	resonance Raman spectroscopy
RS	Raman spectroscopy
SECSY	spin-echo correlated spectroscopy

(continued)

Methods and Techniques In Biophysics and Biochemistry (continued)

SEM	scanning electron microscopy
SERS	surface-enhanced Raman spectroscopy
SEXAF	surface-extended X-ray absorption fine structure (spectroscopy)
SFC	supercritical-fluid chromatography
SOM	scanning optical microscopy
SPA	scintillation proximity assay
SSEE	single-side-band edge enhancement (microscopy)
STM	scanning tunneling (electron) microscopy
TEB	transient electric birefringence
TEM	transmission electron microscopy
TLC	thin-layer chromatography
TMR	topical magnetic resonance
TOCSY	total correlation spectroscopy
TREELS	time-resolved electron energy-loss spectroscopy
TREES	time-resolved europium(III) excitation spectroscopy
TRMS	time-resolved mass spectrometry
TUNEL	terminal deoxynucleotidyl transferase-mediated dUTP-X nick end labeling
UPES	ultraviolet photoelectron spectroscopy
UV	ultraviolet (spectroscopy)
VIS	visible (spectroscopy)
VPC	vapor phase chromatography
VPIR	vapor phase infrared (spectroscopy)
VUV	vacuum ultraviolet (spectroscopy)
XPES	X-ray photoelectron spectroscopy
XRD	X-ray diffraction

Reagents and Technical Terms Related to Methods

A/D	analogue-to-digital
AC, ac	alternating current
AICAR	5-amino-4-imidazolecarboxamide
BAL	British antilewisite
bp	boiling point
BPL	β-propiolactone
BSA	bovine serum albumin
CHAPS	3-[(3-cholamidopropyl)dimethylammonio]-1-propanesulfonate
CM	O-carboxymethyl (cellulose)
CT	charge transfer

(continued)

Reagents and Technical Terms Related to Methods (continued)

D/A	digital-to-analogue
DAPI	4',6-diamidino-2-phenylindole
DC, dc	direct current
DCCD	dicyclohexylcarbodiimide
DCMU	dichloromethylurea
DDT	dichlorodiphenyltrichloroethane
DEAE	O-diethylaminoethyl (cellulose)
DFP	diisopropylfluorophosphate
DMO	5,5-dimethyloxazolidinedione
DMSO	dimethyl sulfoxide
DNP	2,4-dinitrophenol
DOTAP	N-[1-(2,3-dioleoyloxy)propyl]-N,N,N,N-trimethylammonium methylsulfate
DTE	dithioerythritol
DTNB	5,5'-dithiobis(2-nitrobenzoic acid)
DTT	dithiothreitol
ECTEOLA	epichlorhydrin triethanolamine (cellulose)
EDTA	ethylenediaminetetraacetic acid
EGTA	ethylene glycol bis(b-aminoethylether)-N,N,N',N'-tetraacetic acid [*correctly* ethylene bis(oxonitrilo)tetraacetic acid]
EMS	ethylmethane sulfonate
FBS	fetal bovine serum
FID	flame ionization detector
FITC	fluorescein isothiocyanate
G value	number of molecules sensitized by ionizing radiation per 100 eV absorbed
HA	hydroxyapatite
HSA	human serum albumin
IAA	iodoacetic acid
J	coupling constant (in Hz)
laser	light amplification by stimulated emission of radiation
maser	microwave amplification by stimulated emission of radiation
MFR	methanofuran
mp	melting point
MPA	3-mercaptopropionic acid
NEM	N-ethylmaleimide
PABA	p-aminobenzoic acid
PAS	periodic acid Schiff's stain

(continued)

Reagents and Technical Terms Related to Methods (continued)

PCMB	*p*-chloromercuribenzoic acid
PEG	polyethylene glycol
PITC	pyrene isothiocyanate; phenyl isothiocyanate
PMSF	phenylmethylsulfonyl fluoride
POPOP	1,4-bis-2-(5-phenyloxazolyl)benzene
PPO	2,5-diphenyloxazole
PQQ	pyrroloquinoline quinone
R_m	log $(1 - R_F)/R_F$ (in chromatography)
radar	radiowave detection and ranging
RF	rate of movement relative to solvent front
RRT	relative retention time
SDS	sodium dodecyl sulfate
SITS	4-acetamido-4'-isothiocyanostilbene-2,2'-disulfonic acid
TEMED	*N,N,N,N*-tetramethylenediamine
TEMPO	2,2,6,6-tetramethylpiperidine-1-oxyl
TLCK	N-tosyllysine chloromethyl ketone
Tween	polyoxyalkylene sorbitan monoester of fatty acid (a detergent)
UHF	ultrahigh frequency
V_e	elution volume
V_0	void volume
VHF	very high frequency
δ	chemical shift (in ppm; NMR)

Buffers

Commonly used buffers are listed here, such as need not be defined if the abbreviation is used.

Aces	2-[(2-amino-2-oxoethyl)amino]ethanesulfonic acid
Ada	[(carbamoylmethyl)imino]diacetic acid
Bes	2-[bis(2-hydroxyethyl)amino]ethanesulfonic acid
Bicine	*N,N*-bis(2-hydroxyethyl)glycine
Bistris	2-[bis(2-hydroxyethyl)amino]-2-(hydroxymethyl)propane-1,3-diol
Bistris-propane	1,3-bis[tris(hydroxymethyl)methylamino]propane
Chaps	3-[3-cholamidopropyl)dimethylammonio]-1-propanesulfonic acid
Ches	2-(*N*-cyclohexylamino)ethanesulfonic acid
Hepes	4-(2-hydroxyethyl)-1-piperazineethanesulfonic acid
Hepps	4-(2-hydroxyethyl)-1-piperazinepropanesulfonic acid
Mes	4-morpholineethanesulfonic acid
Mops	4-morpholinepropanesulfonic acid

Pipes	1,4-piperazinediethanesulfonic acid
Taps	3-{[2-hydroxy-1,1-bis(hydroxymethyl)ethyl]amino}-1-propane-sulfonic acid
Tes	2-{[2-hydroxy-1,1-bis(hydroxymethyl)ethyl]amino}-1-propanesulfonic acid
Tricine	*N*-[2-hydroxy-1,1-bis(hydroxymethyl)ethyl]glycine
Tris	2-amino-2-hydroxymethylpropane-1,3-diol

Enzyme Kinetics

Name of quantity	Symbol	Units
rate constant		
for first-order reaction	k	s^{-1}
for second-order reaction	k	$mol^{-1}\ L\ s^{-1}$
catalytic constant	k_{cat}	$mol\ L^{-1}\ s^{-1}$
rate of reaction	v	$mol\ L^{-1}\ s^{-1}$
forward and reverse rates of reaction	$k_i,\ k_{-i}$	s^{-1}
specificity constant	k_{cat}/K_m	$mol^{-1}\ L\ s^{-1}$
initial rate of reaction	v_0	$mol\ L^{-1}\ s^{-1}$
limiting (*formerly* maximum) rate of reaction v_{lim} is not official but it is preferable to V_{max}	$V,\ v_{lim}$	$mol\ L^{-1}\ s^{-1}$
maximum rate of reaction	v_{max} (if a true maximum occurs)	$mol\ L^{-1}\ s^{-1}$
flux of S	$J_s = A^{-1}ds/dt$	e.g., $m^{-2}\ mol\ L^{-1}\ s^{-1}$
Michaelis constant	K_m	$mol\ L^{-1}$
half-saturation constant	$[A]_{0.5}$ (in non-Michaelis kinetics, substrate A)	$mol\ L^{-1}$
substrate dissociation constant	K_s	$mol\ L^{-1}$
inhibition constant	K_i	$mol\ L^{-1}$
elasticity coefficient	$\varepsilon_s^v = \partial \ln v\ /\ \partial \ln[S]$	1
flux control coefficient	$C_E^J = (\partial J/J)/(\partial E/E)$	1
Hill coefficient	h	1
enzyme activity	kat (katal)	$mol\ s^{-1}$
		U (unit), e.g., $\mu mol\ min^{-1}$

No attempt has been made to list abbreviations of enzymes, for two reasons: (1) there are now well over 5000 enzyme names (including systematic, abbreviated, and trivial) while an extremely small percentage of them have been given acceptable abbreviations (such as LDH for lactate dehydrogenase or PKC for protein kinase C); (2) in a great many papers on a particular enzyme, an ad hoc abbreviation is coined and used, with no hope of surviving for more than the paper in question (such as FTase for farnesyltransferase or DAD for D-altronate dehydratase). An exception made here are the restriction endonucleases which have acquired more or less sanctioned abbreviations.

Restriction Endonucleases

Restriction enzymes or restriction nucleases are divided into three types: type I (EC 3.1.21.3), type II (EC 3.1.21.4), and type III (EC 3.1.21.5). They are abbreviated with a three-letter acronym in italics (and sometimes with a capital upright letter and a number), plus a roman numeral reflecting the sequence of discovery, e.g., *Eco*R124 II. The following is not an exhaustive list but rather a selection of the typical abbreviations of commonly used enzymes, arranged according to the microorganism from which the given enzyme has been isolated. Most restriction endonucleases are of type II. The ones that are of type I or type III are so described in parentheses.

Acetobacter aceti	*Aat* II
Acetobacter pasteurianus	*Apa* I, *Apa*L I
Acinetobacter calcoaceticus	*Acc* I
Acinetobacter lwofii	*Alw* I, *Alw*26 I, *Alw*N I
Aeromonas hydrophila	*Ahd* I
Agrobacterium gelatinovorum	*Age* I
Anabaena flos-aquae	*Afl* II, *Afl* III
Anabaena variabilis	*Ava* I, *Ava* II, *Avr* II
Aphanothece halophytica	*Aha* III
Aquaspirillum serpens	*Ase* I
Arthrobacter citreus	*Aci* I
Arthrobacter luteus	*Alu* I
Arthrobacter protophormiae	*Apo* I
Bacillus amyloliquefaciens	*Bam*H I
Bacillus aneurinolyticus	*Ban* I, *Ban* II
Bacillus brevis	*Bbv* I
Bacillus caldolyticus	*Bcl* I
Bacillus coagulans	*Bcg* I
Bacillus globigii	*Bgl* I, *Bgl* II
Bacillus laterosporus	*Bbs* I

(continued)

Bacillus lentus	*Blp* I
Bacillus pumilus	*Bpm* I
Bacillus sphaericus	*Bsg* I, *Bsp*1286 I
Bacillus stearothermophilus	*Bsa* I, *Bsa*A I, *Bsa*B I, *Bsa*H I, *Bsa*J I, *Bsa*W I, *Bsi*E I, *Bsi*HKA I, *Bsi*W I, *Bsm* I, *Bsm*B I, *Bsm*F I, *Bso*B I, *Bsr* I, *Bsr*B I, *Bsr*D I, *Bsr*FI, *Bsr*G I, *Bss*H II, *Bss*K I, *Bss*S I, *Bst*B I, *Bst* EII, *Bst*N I, *Bst*U I, *Bst*X I, *Bst*Y I, *Bst*1107 I
Bacillus subtilis	*Bsu*36 I
Bacteroides fragilis	*Bfa* I
Caryophanon latum	*Cla* I
Clostridium acetobutylicum	*Cac*8 I
Corynebacterium sp.	*Csp*6 I
Deinococcus radiodurans	*Drd* I
Deinococcus radiophilus	*Dra* I, *Dra* III
Desulfovibrio desulfuricans	*Dde* I
Enterobacter aerogenes	*Eae* I, *Ear* I
Enterobacter agglomerans	*Eag* I
Enterobacter cloacae	*Ecl*136 II
Erwinia herbicola	*Ehe* I
Escherichia coli	(type I): *Eco*A I, *Eco*B I, *Eco*D I, *Eco*DR2, *Eco*DR3, *Eco*DXX I, *Eco*K I, *Eco*R124 I, *Eco*R124 II, *Eco*RD2, *Eco*RD3, *Eco*prr I, *Sty*LT III, *Sty*SJ, *Sty*SP I, *Sty*SQ; (type II): *Eco47* III, *Eco*57 I, *Eco*N I, *Eco*O109 I, *Eco*R I, *Eco*RV, *Sty* I; (type III); *Eco*P I, *Eco*P151, *Sty*LT I
Fischerella sp.	*Fsp* I
Flavobacterium okeanokoites	*Fok* I
Frankia sp.	*Fse* I
Fusobacterium nucleatum	*Fnu*4H I
Haemophilus parahaemolyticus	*Hph* I
Haemophilus aegypticus	*Hae* II, *Hae* III
Haemophilus haemolyticus	*Hha* I
Haemophilus influenzae	(type II): *Hin*c II, *Hin*d III, *Hin*f I, *Hin*P1 I; (type III): *Hin*f III
Haemophilus parainfluenzae	*Hpa*I, *Hpa*II
Klebsiella pneumoniae	*Kpn* I
Kluyvera ascorbata	*Kas* I
Methanobacterium wolfeii	*Mwo* I

(continued)

Micrococcus luteus	*Mlu* I
Micrococcus sp.	*Msc* I, *Mse* I
Microcoelus sp.	*Mst* I
Moraxella bovis	*Mbo* I, *Mbo* II
Moraxella nonliquefaciens	*Mnl* I
Moraxella osloensis	*Msl* I
Moraxella sp.	*Msp* I, *Msp*A1 I
Mycoplasma fermentans	*Mfe* I
Neisseria cinerea	*Nai* I
Neisseria denitrificans	*Nde* I
Neisseria gonorrhoeae	*Ngo*M I
Neisseria lactamica	*Nla* III, *Nla* IV
Neisseria mucosa heidelbergensis	*Nhe* I
Neisseria sicca	*Nsi* I
Nocardia aerocolonigenes	*Nae* I
Nocardia argentinensis	*Nar* I
Nocardia corallina	*Nao* I
Nocardia otitidis-caviarum	*Not* I
Nocardia rubra	*Nru* I
Plesiomonas shigelloides	*Psh*A I
Proteus vulgaris	*Pvu* I, *Pvu* II
Providencia stuartii	*Pst* I
Psedomonas putida	*Ppu*M I, *Ppu*10 I
Pseudomonas aeruginosa	*Pae*R7 I
Pseudomonas fluorescens	*Pfl*M
Pseudomonas lemoignei	*Ple* I
Pseudomonas maltophila	*Pml* I
Pseudomonas mendocina	*Pme* I
Pseudomonas sp.	*Psp*1406 I
Pseudomonas alcaligenes	*Pas* I
Rhodopseudomonas sphaeroides	*Rsa* I, *Rsr* II
Saccharopolyspora sp.	*Sap* I
Serratia marcescens	*Sma* I
Sphaerotilus sp.	*Spe* I, *Spe* II
Staphylococcus aureus	*Sau*3A I, *Sau*96 I
Streptococcus cremoris	*Scr*F1
Streptococcus faecalis	*Sfa*N I
Streptococcus faecium	*Sfc* I
Streptomyces achromogenes	*Sac* I, *Sac* II
Streptomyces albus	*Sal* I
Streptomyces caespitosus	*Sca* I

(continued)

Streptomyces fimbriatus	*Sfi* I
Streptomyces phaeochromogenes	*Sph* I
Streptomyces tubercidicus	*Stu* I
Thermus aquaticus	*Tai* I
Thermus filiformis	*Tfi* I
Thermus sp.	*Tse* I, *Tsp*45, *Tsp*509 I, *Tsp*R I, *Tsp*E I, *Tsp*R I
Thermus thermophilus	*Tth*111 I, *Tth*111 II
Xanthomonas badrii	*Xba* I
Xanthomonas campestris	*Xcm* I
Xanthomonas holcicola	*Xho* I, *Xho* II
Xanthomonas malvacearum	*Xma* I, *Xma* III
Xanthomonas manihotis	*Xmn* I

What with the appearance of repetitive modules in protein structure, quite frequently in extracellular segments of proteins, there are attempts to assign abbreviations to them, such as those collected by P. Bork of Heidelberg and A. Bairoch of Geneva and published in 1995 by *Trends in Biochemical Sciences*. I list here a sampling.

CA	cadherin	NL	notch/lin-12
CCP, CP	complement control protein	PD	P-type (trefoil)
CLECT, CTL	C-type lectin	SP	somatomedin B
F1, F2, F3	fibronectin type I, II, III	SR	scavenger receptor Cys-rich
HX	hemopexin-like	TY	thyroglobulin type-1
KR	kringle		

Catalytic domains in enzymes are designated with capital letters starting with R, S, T,…, regulatory domains with lowercase letters. starting with a, b, c,…

Cell Biology

A, B, D cells	cells of the pancreas
A-band	anisotropic part of the sarcomere
ABP	actin-binding proteins
ADF	actin-depolymerizing factor
BHK (cells)	baby hamster kidney cells
BMP	bone morphogenetic protein
BFU-E	burst-forming units (of erythrocytes)
C-proteins	thick-filament-associated proteins in striated muscle
CDC	cell division cycle
CGN	cis-Golgi network
CHF	chick heart fibroblast
CHO (cells)	fibroblast line from Chinese hamster ovary
CM	cytoplasmic membrane, cell membrane
CURL	compartment for uncoupling of receptors and ligands
DC	dendritic cell
E-face	inner face of the membrane lipid outer layer (freeze fracturing)
EHS	Englebreth-Holm-Swarm sarcoma cells
I-band	isotropic band of striated muscle sarcomere
EB	elementary body (of *Chlamydia*)
ECM	extracellular matrix
ER	endoplasmic reticulum
FA	focal adhesion
FN	fibronectin
G_1 (phase)	the gap period of cell cycle between cell division and S phase
G_2 (phase	the gap period of cell cycle between S phase and M phase
GA	Golgi apparatus
GERL	Golgi-associated endoplasmic reticulum, forming lysosomes
HC	heavy chain
HeLa (cells)	human epitheliual cell line obtained from Henrietta Lacks
IC	intermediate chain
IF, IMF	intermediate filament
INM	inner nuclear membrane
L, M, P, S rings	parts of bacterial flagellum
LC	light chain

(continued)

M (phase)	cell cycle period when mitosis takes place
M-band	central region of A-band in striated muscle sarcomere
MCP	methyl-accepting chemotaxis protein
MPF	M-phase-promoting factor
MF	microfilament
MT	microtubule
MTOC	microtubule-organizing center
N-lines	regions in striated muscle sarcomere
NE	nuclear envelope
NPC	nuclear pore complex
OM	outer membrane
ONM	outer nuclear membrane
P-face	inner face of the membrane lipid inner layer (freeze-fracturing)
R point	a point in the late G_1 phase of cell cycle
RER	rough endoplasmic reticulum
RES	reticuloendothelial system
ROS	rod outer segment
RS	radial spoke
S (phase)	cell cycle period when DNA is replicated or synthesized
SER	smooth endoplasmic reticulum
SF	stress fiber
SPB	spindle-pole body
TF	transverse filament
TGN	trans-Golgi network
Z-disc (= Z-line)	region of striated muscle sarcomere

Genetics

Abbreviations and symbols used in genetics as a whole lack uniformity and very few, if any, conventions used in bacterial genetics would apply to the fruit fly or thale cress. First to be treated here are abbreviations applying to chromosomes, followed by abbreviations applying to genes and their expression products.

In classical genetics, P stands for parental generation, F_1, F_2, etc., for consecutive filial generations, B_1, B_2 for backcross generations, I_1 and I_2 for inbred generations, n for the haploid number of chromosomes.

Chromosomes

These are designated throughout the various categories of multicellular organisms by upright arabic numerals and/or upright capital letters. For yeasts and fungi, upright roman numerals are used; examples follow:

Organism	Chromosomes (haploid + sex)
Caenorhabditis elegans (a nematode)	I-V, X
Drosophila melanogaster (fruit fly)	*1, 2, 3, 4, X*
Bombyx mori (silkworm)	1-27, X, Y
Mus sp. (mouse)	1-19, X, Y
Bos taurus (cattle)	1-29, X, Y
Felis catus (cat)	A1, A2, A3, B1, B3, B4, C1, C2, D1, D2, D3, D4, E1, E2, E3, F1, F2, X, Y
Capra hircus (goat)	1-29, X, Y
Equus caballus (horse)	1-31, X, Y
Sus scrofa (pig)	1-18, X, Y
Oryctolagus cuniculus (rabbit)	1-21, X, Y
Ovis aries (sheep)	1-26, X, Y
Homo sapiens (human)	1-22, X, Y
Gossypium sp. (cotton)	1-26
Allium sp. (onion)	1C-8C
Zea mays (corn)	1-10
Oryza sativa (rice)	1-12
Secale sp. (rye)	1R-7R

(continued)

Lycopersicum esculentum (tomato)	1-2
Triticum sp. (wheat)	1-21
Avena sp. (oat)	1-21
Saccharomyces cerevisiae (baker's yeast)	I-XVI
Aspergillus nidulans	I-VIII
Neurospora crassa	I-VII
Schizosaccharomyces pombe	I-III

Chromosomal aberrations and anomalies are designated as follows:

Del	deletion	R	rings
Df	deficiency, deletion	Rb	Robertsonian translocation
Dp	duplication	T	translocation
In	inversion	Tp	transposition
Is	insertion	Ts	trisomy
Ms	monosomy		

If these abbreviations occur in *Caenorhabditis elegans*, the fruit fly, cotton, rice, soybean, and wheat, they are italicized. This is an unwarranted usage and one may wonder whether a chromosomal aberration studied in a new species will be in roman type or in italics.

There are additional symbols and abbreviations relating to human cytogenetics:

AI	1st meiotic anaphase	mar	marker chromosome
AII	2nd meiotic anaphase	mat	maternal origin
ace	acentric fragment	mos	mosaic
b	break	oom	oogonial metaphase
cen	centromere	p	short arm of chromosome
chi	chimera	PI	first meiotic prophase
cs	chromosome	PII	second meiotic prophase
ct	chromatid	pac	pachytene
cx	complex	pat	paternal origin
del	deletion	prx	proximal
dic	dicentric	q	long arm of chromosome
dip	diplotene	r	ring chromosome
dir	direct	rec	reciprocal
dis	distal	rea	rearrangement
dit	dictyate	rec	recombinant chromosome
dup	duplication	s	satellite
e	exchange	sce	sister chromatid exchange
end	endoreduplication	sdl	sideline, subline

(continued)

f	fragment	sl	stemline
fem	female	spm	spermatogonial metaphase
g	gap	t	translocation
h	secondary constriction	tan	tandem
i	isochromosome	ter	terminal end of chromosome
ins	insertion	tr	triradial
inv	inversion	tri	tricentric
lep	leptotene	var	variable chromosome region
MI	1st meiotic metaphase	xma	chiasma
MII	2nd meiotic metaphase	zyg	zygotene
mal	male		

The type of DNA sequence in a chromosome, if it has no recognized function, is designated with an upright letter: S, F, Z, E, X, Y.

Genes and Their Expression Products

A single general rule is that gene and allele abbreviations are set in italics while their phenotypes and products of their expression (different RNAs and proteins) are set in roman type.

Prokaryotes

Genotype designations must use a three-letter symbol written in italics: *his, ara, gln.* If there are several loci coding for functionally related proteins, they are designated with an italicized capital letter: *araA, araB.*

Wild-type alleles may be designated with a superscript sign: *his*+. Superscript minus signs are not used to indicate a mutant locus. Designations of amber (Am), temperature-sensitive (Ts), constitutive (Con), and cold-sensitive (Cs) mutations, and production of a hybrid protein (Hyb) should follow the allele designation in parentheses: *hisD21*(Ts).

To describe a promoter site in the gene a *p* is added, for termination site a *t*, for operator site an *o*, and for attenuation site an *a*: *lacZp, lacZt, lacZo, lacZa.* Subscripts may only be used to distinguish between identically named genes from different organisms: *his*$_{K-12}$ for the K-12-derived *his* gene placed in another organism.

Plasmids are designated usually with three-letter-plus-numeral abbreviations, written in roman letters, e.g., pMB1, ColE1, IncIa. For newly described plasmids the prefixed p is obligatory.

Phenotype designations are made in roman type with the first letter capitalized: Met, with Met+ for a wild-type phenotype and Met- for a mutant phenotype. For further specifying a phenotype, superscripts may be used: Str^s for streptomycin sensitivity.

The proteins coded by a particular gene, unless a specific name for them exists, are designated with upright letters: His2.

Bacteriophage genes are designated with symbols in italics, e.g., *N*, *cI*, *int*, *32*, but their phenotypes and products are roman, e.g., Int; numerical abbreviations of products are preceded by gp, e.g., gp32.

Eukaryotic Microorganisms

The practice in genetic work with yeasts and fungi follows partly that with bacteria in employing italicized, usually three-letter, symbols for the designation of genes, including numbers that refer to the sequence of discovery. The following table reviews the practice in gene and allele designation in various yeasts and fungi, demonstrating, at the same time, the deplorable lack of uniformity in the field.

In general, mitochondrial gene abbreviations should be enclosed in brackets: [*oli A1*].

Feature	Saccharomyces cerevisiae	Schizosaccharomyces pombe	Fusarium sp.	Aspergillus nidulans[a]	Neurospora crassa[b]
Gene or locus	ade5	arg1	met, MET	arg	suc, Bml
Dominant allele	CUP1	leu2	MET	Arg, arg	Sk
Recessive allele	arg2	leu	met	arg	suc
Wild-type gene	ARG2+	arg1+	Met+	argA+	leu-2+
Gene conferring a specific property	con^{r}1	—	CUP^{R}2	argA^{d}12	argC^{R}2
Phenotype					
nonrequiring a nutrient	Arg$^-$	leu$^-$	Met$^-$	Arg$^-$	ad$^-$
requiring a nutrient	Arg$^+$	leu$^+$	Met$^+$	Arg$^+$	ad$^+$
Protein product	Gal2, Gal2p	leu2, leu2p	Met	Arg	Suc

[a]Until recently, 1–5 letters were used to designate the genes.
[b]The gene designations consist of 1–4 letters, such as *fr*, *suc*, *acr-3*, *ta*, *os-4*, *phen-1*, *mo(P1417)* and *smco-5*.

Plants

In classical as well as in modern genetics of plants few rules are available as to gene designation. Although always in italics, the symbols may include (in tomato) *y*, *co*, *spa*, *Hr*, *Cf-2*, *rust*, *pca-2*, *v2*, and *ms9*, or (in corn) *as*, *d1*, *ra2*, *Cg*, *Rf1*, *rgd*, and *A2*, or (in rice) *wx*, *d*, *Pgi*, or (in oat) *sc*, *Tg*, *cda*, or (in *Arabidopsis thaliana*) *EMB1*, *DET*, *DIS*.

A systematic policy now exists for naming sequenced plant genes. Abbreviations for genetic loci in plants (called *mnemonic* designation by the International Society for Plant Molecular Biology) consist mostly of three italicized letters, followed by an italicized number, such as *Chi2* (for chitinase, class II); occasionally, four-letter abbreviations are used, such as *Lhcb1* (for light-harvesting complex II protein of type I). In *Arabidopsis thaliana*, the practice now observed is like with *Saccharomyces cerevisiae*, but phenotype designations may but need not use superscripts (e.g., Abc+, Abc-). Designations of genes residing in the nucleus begin with a capital letter (as above), of those residing in mitochondria or chloroplasts proceed from the "bacterial" usage and are all in lowercase letters, such as *atpB* (for chloroplast ATP synthase, β subunit).

It is recommended in general that abbreviated gene designations only be permitted if the protein product coded by the gene has been identified, i.e., not for any phenotypic trait, as the case is with bacteria. Phenotypes are designated with roman letters: Arg.

In all plants a dominant allele should be designated with an initial capital letter: *Dek 3*; a recessive allele in lowercase letters: *dek 12*. In many cases, special abbreviations are used for modulatory agents, viz. *I*, *Su*, *En*, and *M* for, respectively, inhibitors, suppressors, enhancers, and modifiers of dominant alleles; with recessive alleles, the abbreviations are all lowercase.

The prevailing practice in naming the gene products is in upright capital letters: DET1, ADH1.

Animals

The only rule here is that the gene abbreviation is italicized. Otherwise, there is a great variability in the number of letters, subscripts, superscripts, etc. Thus, in the mouse one encounters *s*, *du*, *stb*, *N*, *Sd* , *Sha*, *Es-3*; in chicken *h*, *P*, *fr*, *Li*, s^{al}, K^N; in fruit fly *y*, *N*, *vt*, *Cb*, *fla*, *Bxd*, *Antp*, *E(s)*, *M(1)4BC*, *l(1)Q214*, *Su(S)*, *Stp-2*, $su(w^{ch})$; in silkworm: *e*, *L*, *os*, *Ze*, *mal*, *Spc*, *w-3*, p^B, $+^{p4}$; in fish: *IDHP**, *LDH.A**. In the recent object of investigation, the nematode *Caenorhabditis elegans*, three-letter symbols are used consistently, e.g., *dpy-5*, *let-37*.

Phenotypes in *C. elegans* are abbreviated with roman letters (Unc, Dpy), gene products in roman capital letters (UNC-13). In *D. melanogaster*, phenotypes are roman-type equivalents of the gene designation, e.g., the genotype *white* determines the phenotype white. The gene products are designated with uppercase letters, in italics if mRNA (*AL*), roman if protein (AL) for products of the *al* (or *aristaless*) gene.

In chicken, the phenotypes are designated by the same symbol as the gene but ar in upright (roman) letters, just like the protein products. In mouse, the phenotypes are all uppercase and upright; the same holds for gene products.

Symbols for oncogenes are based on the retrovirus where the oncogene was first identified. The prefix lowercase letter refers to the homolog type (c-*myc* for cellular, from myelocytomatosis); capital letters indicate cellular localization (N-*erb A*, for nuclear, from avian erythroblastosis).

In modern human genetics all gene and allele designations are in capital italics (*ACADS*, *G6PD*, *HBB@*) while the protein products, like in all organisms without exception, are upright and, like with plants but not microorganisms, are all in capitals (PDM1).

Taxonomy

Although hardly a field of life sciences in the usual sense of the word, taxonomy is an important scientific discipline that cuts across all the areas where living organisms are described, i.e., from virology to bacteriology to botany to zoology.

Some abbreviations unique to taxonomy are listed here first. They include those used in numerical taxonomy.

subg.	subgenus	f.sp.	special form
sect.	section	nom.cons.	name to be kept
subsect.	subsection	nom. dub.	dubious name
ser.	series	nom. rejic.	name to be rejected
subser.	subseries	nom. rev.	revived name
sp.	species	nom. conf.	confused
ssp., subsp.	subspecies	nom. legit.	legitimate name
var.	variety	nom. ambig.	ambiguous name
r.	subvariety	OTU	operational taxonomic name
bv.	biovar	S_J	Jaccard coefficient
cv.	cultivar	S_{SM}	simple matching coefficient
pv.	pathovar	S_Y	Yule coefficient
sv.	serovar	S (matrix)	simulating
f.	form	CD	coefficient of divergence

Tables of hierarchical divisions in the various systematic fields, such as are characterized by well-established endings of their designations, follow.

Suffixes for Taxonomical Categories

Taxon	Bacteria	Example	Algae	Example	Higher plants	Example
Division	-phyta	Bacteriophyta	-phyta	Chromophyta	-phyta	Spermatophyta
Subdivision	—	—	—	—	-phytina	Angiospermophytina
Class	-phyceae	Bacteriophyceae	-phyceae	Phaeophyceae	-opsida, -atae	Magnoliatae
Subclass	—	—	-phycidae	—	-idae	Asteridae
Superorder	—	—	-atae	Isogeneratae	-anae	Lamianae
Order	-ales	Pseudomonadales	-ales	Dictyotales	-ales	Lamiales
Suborder	-ineae	Pseudomonadineae	-ineae	—	-ineae	—
Family	-aceae	Pseudomonadaceae	-aceae	Dictyotacae	-aceae	Laminaceae
Subfamily	-oideae	Pseudomonadoideae	-oideae	—	-oideae	Laminoideae
Tribe	-eae	Pseudomonadeae	-eae	—	-eae	Saturajeae
Subtribe	-inae	Pseudomonadinae	-inae	—	-inae	Thyminae

Virus taxonomy only begins at the level of families where the ending is -viridae, e.g., Herpesviridae. For subfamilies the ending is -virinae, e.g., Alphavirinae.

Some schools of bacterial taxonomy use italics for all taxa, a practice that is inconsistent with the usage in all other categories of organisms.

74

Taxon	Fungi	Example	Protozoa	Example	Animals	Example
Phylum	-mycota	Eumycota	-a	Sarcomastigophora	-a, -es. -i	Chordata
Subphylum	-mycotina	Eumycotina	-a	Sarcodina	-a	Vertebrata
Superclass	-a	–	-a	Rhizopoda	-a	Gnathostomata
Class	-mycetes	Basidiomycetes	-ea	Lobosea	-a, -es	Mammalia
Subclass	-mycetidae	Holobasidiomycetidae	-ia	–	-a, -es, -i	Theria
Superorder	-mycetales	Hymenomycetales	-idea	–	-ii, -ei, -ae, -ia	Eutheria
Order	-ales	Agaricales	-ida	Amoebida	-a, -formes, -i	Primates
Suborder	-ineae	–	-ina	Amoebina	–	Anthropoidea
Superfamily	–	–	-oidea	Amoeboidea	-oidea	Hominoidea
Family	-aceae	Coprinaceae	-idae	Amoebidae	-idae	Hominidae
Subfamily	-oideae	Coprinoideae	-inae	–	-inae	Homininae
Tribe	-eae	–	-ini	–	–	–
Subtribe	-inae	–	–	–	-ina	–

Some classes of Protozoa are grouped with algae and then they carry designations with suffixes corresponding to that group or organisms.

All the taxa listed in the tables are to be set in roman (upright) type. Starting with the genus, italics should be used, thus *Tussilago farfara* (genus and species of colts-foot). This holds also for subspecies, thus *Veronica hederifolia* ssp. *lucorum* (lilac ivy-leaved speedwell). However, with cultivated varieties or races of plants, so-called cultivars, the custom is to use roman letters, starting with capitals, thus *Clematis florida*, cv. Duchess of Edinburgh.

In microbiology, names of all taxa ordered below species, viz. subspecies. variety, subvariety, form, and special form, are written in italics. Only the very lowest taxon, (physiological) race, is written in upright letters.

Virology

AAV	adeno-associated virus
AcNPV	*Autographa californica* nuclear polyhedrosis virus
ABDV	*Arrhenatherum* blue dwarf virus
AbMuLV	Abelson murine leukemia virus
AEV	avian erythroblastosis virus
AKV	a murine leukemia virus
ALV	avian leukosis virus
AMV	avian myeloblastosis virus
AmuLV	Abelson murine leukemia virus
ARV	AIDS-associated retrovirus
ASBV	avocado sunblotch viroid
ASFV	African swine fever virus
ASV	avian sarcoma virus
BBV	black beetle virus
BCPV	bovine cutaneous papilloma virus
BCTV	beet curly top virus
BGAV	blue-green algal virus
BGMV	bean golden mosaic virus
BLV	bovine leukosis virus
BMV	brome mosaic virus
BPPV	bovine paragenital papilloma virus
BPV	bovine papilloma virus
BSV	burdock stunt viroid
BTV	blue tongue virus
BWYV	beet western yellows virus
BYDV	barley yellow dwarf virus
CAEV	caprine arthritis-encephalitis virus
CaMV	cauliflower mosaic virus
CAV	croup-associated virus
CCCV	coconut cadang-cadang viroid
CCMV	chrysanthemum chlorotic mottle viroid
CCV	channel catfish virus; canine coronavirus
CDV	Carr's disease virus
CEV	*Citrus* exocortis viroid
CGMMV	cucumber green mottle mosaic virus

(continued)

CHV	canine herpes virus
CLMV	cauliflower mosaic virus
CLV	cassava latent virus
CMV	cytomegalovirus; cucumber mosaic virus
CPFV	cucumber pale fruit viroid
CPMV	cowpea mosaic virus
CPV	cytoplasmic polyhedrosis virus
CRHV	cottontail rabbit herpes virus
CrPV	cricket paralysis virus
CSMV	*Chloris* striate mosaic virus
CSV	chrysanthemum stunt viroid; chick syncytial virus
CTDV	cereal tillering disease virus
DHBV	duck hepatitis B virus
DKV	deer kidney virus
DLV	defective leukemia virus
DNV	densovirus
EAV	equine abortion virus
EBV	Epstein-Barr virus
ECDOV	enteric cytopathogenic dog orphan virus
ECHOV	enteric cytopathogenic human orphan virus
ECMOV	enteric cytopathogenic monkey orphan virus
ECPOV	enteric cytopathogenic porcine orphan virus
EHDV	epizootic hemorrhagic disease virus
EHV	equine herpes virus
EIAV	equine infectious anemia virus
EPV	entomopox virus
ERV	equine rhinopneumonitis virus
FCV	feline colicivirus
FDV	Fiji disease virus
FeLV	feline leukemia virus
FeSV	feline sarcoma virus
FHV	flock house virus
FLV	Friend leukemia virus
FMDV	foot-and-mouth disease virus
FPV	fowl pest virus; fowl plague virus
FV	Friend virus
GALV	Gibbon ape leukemia virus
GSHV	ground squirrel hepatitis virus
GV	granulosis virus; Gross virus
HaMuSV	Harvey murine sarcoma virus
HAV	hepatitis A virus; hemadsorption virus

(continued)

HBV	hepatitis B virus
HCMV	human cytomegalovirus
HEV	hemagglutination encephalomyelitis virus; hepatoencephalitis virus
HHDV	human hepatitis D virus
HIV	human immunodeficiency virus
HPV	human parvovirus; human papilloma virus
HRV	human rhinovirus
HSV	herpes simplex virus; hop stunt viroid
HTLV	human T-cell lymphoma virus
HVH	herpes virus hominis
HVP	herpes virus Papio
IBV	infectious bronchitis virus
IEEHV	immediate early equine herpes virus
IPNV	infectious pancreatic necrosis virus
JCV	Jamestown Canyon virus
JSV	Jerry-Slough virus
KERV	Kentucky equine respiratory virus
KMSV	Kirsten murine sarcoma virus
LAHV	leukocyte-associated herpes virus
LAV	lymphoadenopathy-associated virus; leafhopper A virus
LCV	lymphocytic choriomeningitis virus
LEPV	low egg passage virus
LEV	*Lolium* enation virus
LLV	lymphatic leukemia virus
LPV	lymphotropic papovavirus
MCDV	maize chlorotic dwarf virus
MCFV	mink cell focus-forming virus
MCMV	murine cytomegalovirus
MDMV	maize dwarf mosaic virus
MDV	Marek's disease virus; mucosal disease virus
MECV	murine encephalomyocarditis virus
MEV	murine erythroblastosis virus
MHV	Mill Hill virus; murine hepatitis virus
MLV	murine leukemia virus; Maloney leukemia virus
MMSV	mouse Maloney sarcoma and leukemia virus
MMTV	mouse mammary tumor virus
MoMuLV	Moloney murine leukemia virus
MPMV	Mason-Pfizer monkey virus
MRDV	maize rough dwarf virus
MSV	maize streak virus; Moloney sarcoma virus; murine sarcoma virus
MuLV	murine leukemia virus

(continued)

MuMTV	murine mammary tumor virus
MVM	minute virus of mice
NDV	Newcastle disease virus
NEEEV	Near East equine encephalomyelitis virus
NOV	nonoccludid virus
NPV	nuclear polyhedrosis virus
OSDV	oat sterile dwarf virus
POWV	Powassan encephalitis virus
PLRV	potato leaf roll virus
PRCNV	porcine respiratory coronavirus
PSTV	potato spindle tuber viroid
PSV	pangola stunt virus
PTLV	primate T-cell virus
PVH	papilloma virus hominis
PVM	pneumonia virus of mice
PVX	potato virus X
PVY	potato virus Y
RAV	Rous-associated virus
RBSDV	rice black-streaked dwarf virus
REV	reticuloendotheliosis virus
RKV	rabbit polyoma virus
RLV	Rauscher leukemia virus
RMuLV	Rauscher murine leukemia virus
RSV	Rous sarcoma virus; respiratory syncytial virus
ScV	*Saccharomyces cerevisiae* virus
SFV	Semliki forest virus; shipping fever virus; squirrel fibroma virus
SINV	Sindbis virus
SIV	simian immunodeficiency virus
SMuLV	Scripps murine leukemia virus
SRV	Schmidt-Ruppin virus; Shark River virus
SSAV	simian sarcoma-associated virus
SSHV	snow-shoe hare virus
SSV	Schoolman-Schwartz virus; simian sarcoma virus
STNV	satellite tobacco necrosis virus
SVDV	swine vesicular disease virus
SV40	Simian virus 40
TASV	tomato apical stunt viroid
TBEV	tick-borne encephalitis virus
TBRV	tomato black ring virus
TBSV	tomato bushy stunt virus
TBTV	tomato bunchy top viroid

(continued)

TCCAV	transitional-cell cancer-associated virus
TCV	turnip crinkle virus
TGMV	tomato golden mosaic virus
TIV	Tipula iridescent virus
TMEV	Theiler's murine encephalomyelitis virus
TNV	tobacco necrosis virus
TobRV	tobacco ringspot virus
TPMV	tomato "planta macho" viroid
TRV	tobacco rattle virus; Tupaia retrovirus
TSWV	tobacco spotted wilt virus
TTNV	tomato top necrosis virus
TYDV	tobacco yellow dwarf virus
TYMV	turnip yellow mosaic virus
TYNV	tomato yellow net virus
VLP	virus-like particle
VSV	vesicular stomatitis virus
VTMoV	velvet tobacco mottle virus
VZV	varicella-zoster virus
WHV	woodchuck hepatitis virus
WNV	West Nile virus
WTV	wound tumor virus

Microbiology

BOD	biological (biochemical) oxygen demand
CFU, cfu	colony-forming units
ColA, ColB. etc.	colicin A, colicin B, etc.
COD	chemical oxygen demand
CSF	colony-stimulating factor

[Cultivation media are abbreviated with upright capital letters, e.g., CLED = cystine-lactose-electrolyte-deficient medium.]

D	dilution rate
D value (D_{10} value)	decimal reduction time
D_c	critical dilution rate
F factor (F plasmid)	a conjugative plasmid
F-factor	plasmid conferring the ability to conjugate (for fertility)
Hfr (state)	high frequency of recombination
HSP	heat-shock protein
m	maintenance coefficient (s^{-1} or, more often, h^{-1})
MLO	mycoplasma-like organism
PFU	plaque-forming units
PPLO	pleuropneumonia-like organisms
q	specific mass metabolic rate (s^{-1}, h^{-1})
Q_{O2}, Q_{CO2}	metabolic quotients
r	productivity (m^{-3} kg h^{-1})
RQ	respiratory quotient
SCP	single-cell protein
t_d	doubling time (s, min, h, d)
w	mutation rate (s^{-1}, h^{-1})
Y	yield coefficient
η	dilution coefficient (concentration exponent) in disinfection
μ	specific growth rate (s^{-1} or, more often, h^{-1})

Plant Physiology

A number of symbols derived from physics and chemistry are used in this area and these will be listed first, to be followed by abbreviations.

Quantity	Symbol	Units
water potential		
specific	ψ_m	$J\,kg^{-1}$
volumetric	ψ_V	$J\,m^{-3}$, Pa
weight	ψ_f	$J\,N^{-1}$, m
molar	ψ_n	$J\,mol^{-1}$
hydrostatic pressure	P, ψ_P	Pa
osmotic potential	ψ_s, ψ_π	Pa
osmotic pressure	$\pi = -\psi_s$	Pa
volume flux	J_V	$m\,s^{-1}$
solute flux	J_S	$mol\,m^{-1}\,s^{-1}$
solute permeability	P_S	$m\,s^{-1}$
hydraulic conductance	L_p	$m\,s^{-1}\,Pa^{-1}$
(Conductivity should not be used.)		
diffusivity (diffusion coefficient)	D	$m^2\,s^{-1}$
coefficient of heat convection	h_c	$W\,m^{-2}\,K^{-1}$
metabolic heat turnover	M	$J\,kg^{-1}$
photosynthetic irradiance	PI	$W\,m^{-2}$
photosynthetic photon flux	PPF, E_p	$mol\,m^{-2}s^{-1}$
relative humidity	RH	%
solar irradiance incident at surface	I_s	$W\,m^{-2}$
quantum yield	σ	1
specific leaf area	SLA	$m^2\,kg^{-1}$
percent leaf mass	PLM	%
net assimilation rate	NAR	$g\,m^{-2}\,d^{-1}$
relative growth rate	RGR	$g\,kg^{-1}\,d^{-1}$

C3 (plants)	plants fixing CO_2 by the Calvin cycle at ribulose 1,5-bisphosphate
C4 (plants)	plants fixing CO_2 by the Hatch-Slack-Kortschak pathway at phosphoenolpyruvate
CAM (plants)	plants displaying crassulacean acid metabolism
CK	cytokinin

85

CT   circadian time
DD   continuous darkness
GA   gibberellin
IAA   indole-3-acetic acid
LD   light-dark cycle
LL   continuous light
PIIF   proteinase-inhibitor-inducing factor
Ti   tumor-inducing (plasmid) from *Agrobacterium tumefaciens*
PQ   plastoquinone
PSI, PSII   photosystems I and II

Animal Physiology

Most of what is presented in the sections on physics and chemistry applies here. Unlike in many other fields of life sciences, there appears to have been no serious attempt to introduce unified abbreviations in this area. Although some of the leading journals recommend some abbreviations to be used without explanation, this is certainly not a trend to be followed.

ACE	angiotensin-converting enzyme
ACTH	adrenocorticotrophic hormone, corticotropin
ADH	antidiuretic hormone (vasopressin)
AHF	antihemophilic factor
AKH	adipokinetic hormone
ANP	atrial natriuretic peptide
APP	amyloid precursor protein
BDNF	brain-derived neurotrophic factor
BNP	brain natriuretic peptide
CG	choriogonadotropin
CGH	chorionic gonadotrophic hormone
CLSH	corpus luteum stimulating hormone
CNTF	chick neurotropic factor; ciliary neurotropic factor
CRF	coagulase-reactive factor; corticotropin-releasing factor (liberin)
CSF	cerebrospinal fluid; colony-stimulating factor
CSIF	cytokine-synthesis-inhibiting factor
CT	calcitonin
CVF	cobra venom factor
EDRF	epithelium-derived relaxation factor
EGF	epidermal growth factor (urogastrone)
EMM	Eagle's minimal medium
EPF	epidermal proliferation factor; early pregnancy factor
Epi	epithelial
EPSP	excitatory postsynaptic potential
ERP	early receptor potential
ET-1, ET-2, etc.	endothelins
f	respiratory frequency
f_a	fraction of administered dose absorbed

(continued)

F	blood flow into an organ (mL g^{-1} min^{-1})
FGF	fibroblast growth factor
Fib	fibroblastic
FRF	FSH-releasing factor
FSH	follicle-stimulating hormone
GDNF	glial-cell-line-derived neurotrophic factor
GEF	glycosylation-enhancing factor
GFR	glomerular filtration rate
GGF	glial growth factor
GH	growth hormone (somatotropin)
GHRH	somatostatin-releasing hormone
GIF	glycosylation-inducing factor
GIP	gastric inhibitory peptide; glucose-dependent insulinotropic polypeptide
GMF	glial maturation factor
GnRF	gonadotropin-releasing factor
GRP	gastrin-releasing peptide
GSF	glial stimulating factor
HGF	hepatocyte-growth factor
HSP, Hsp	heat-shock-protein
IGF	insulin-like growth factor
KGF	keratinocyte growth factor
LAF	leukocyte-activating factor
LH	luteinizing hormone
LHRF	luteinizing-hormone releasing factor (luliberin)
LIF	leukemia-inhibiting factor
LPH	lipotropic hormone
MEF	migration enhancement factor
MIF	migration-inhibiting factor
MIH	melanocyte-inhibiting hormone
MSH	melanocyte-stimulating hormone (melanotropin)
MSHRH	melanocyte-stimulating-hormone regulating hormone
NAF	neutrophil-activating factor
NGF	nerve growth factor
OT	oxytocin
PA	plasminogen activator
PAF	platelet-activating factor
PDGF	platelet-derived growth factor
PIF	prolactin.inhibiting factor
POMC	proopiomelanocortin
PRH	prolactin-releasing hormone

(continued)

PRL	prolactin
PrP	prion protein
PTH	parathyroid hormone
SP	substance P
SRIH	somatotropin-inhibitory hormone (somatostatin)
T4, Thx	thyroxine
TAF	tumor angiogenesis factor
TF	transfer factor
TGF	transforming growth factor
TNF	tumor necrosis factor
TPF	tissue plasminogen activator
TRH	thyrotropic-hormone releasing hormone (thyrotropin)
TSH	thyroid-stimulating hormone
VIP	vasoactive intestinal peptide
VP	vasopressin

Pharmacology

Many of the symbols and abbreviations used in pharmacology are listed under Biochemistry and Molecular Biology, and under Animal Physiology and Medicine. Only some of the more unique ones are mentioned here.

A	agonist
A^{ss}	amount of drug in body at steady state
ADI	acceptable daily intake
B	antagonist
e	efficacy
EC_{50}	molar (effective) concentration producing 50% of limiting response
ED_{50}	effective dose producing 50% of limiting response
HD_{50}	hemolytic dose which hemolyzes 50% of a blood cell population
HQ	hazard quotient
IC_{50}	molar inhibitor/antagonist dose reducing a control response by 50%
ID_{50}	infectious dose which infects 50% of the population
L	ligand
LD	lethal dose
LOEL	lowest-observed-effect level
MAC	maximum allowable concentration
MBC	minimum bactericidal concentration
MEL	maximum exposure limit
MIC	minimum inhibitory concentration
MLD	minimum lethal dose
MOS	margin of safety
MRL	maximum residue limit
NC_{50}	median narcotic concentration
NOEL	no-observed-effect level
P	partial agonist
p_{LR}	proportion of receptors occupied by ligand L
p_R	proportion of free receptors
R	receptor
TDI	tolerable daily intake
TLV	threshold limit value
a	intrinsic activity
ε	intrinsic efficacy

Receptors for drugs and receptor channels are now receiving consistent abbreviations, such as the following:

A_1, A_{2A}, A_{2B}, A_3	adenosine receptors
AMPA	receptor for α-amino-3-hydroxy-5-methyl-4-isoxazole-propionic acid
ANP_A, ANP_B	atrial natriuretic peptide receptors
AT_1, AT_2	angiotensin receptors
α_{1A}, α_{1B}, α_{1D}	α_1-adrenoreceptors
α_{2A}, α_{2B}, α_{2C}	α_2-adrenoreceptors
BLT, $CysLT_1$, $CysLT_2$	leukotriene receptors
B_1, B_2	bradykinin receptors
BB_1, BB_2, bb_3	bombesin receptors
β_1, β_2, β_3	β-adrenoreceptors
CB_1, CB_2	cannabinoid receptors
CCK_A, CCK_B	cholecystokinin and gastrin receptor
CCK_A, CCK_B	cholecystokinin and gastrin receptors
CCR_1-CCR_5	β-chemokine receptors
CGRP	calcitonin-gene-related peptide receptor
CRF_1, $CRF_{2\alpha}$, $CRF_{2\beta}$	corticotropin-releasing factor receptor
$CXCR_1$-$CXCR_4$	α-chemokine receptors
D_1-D_5	dopamine receptors
DP, FP, IP, TP, EP_1-EP_4	prostanoid receptors
ET_A, ET_B	endothelin receptors
$GABA_1$, $GABA_2$	receptors for γ-aminobutyric acid
GN-1	galanin receptor
H_1, H_2, H_3	histamine receptors
5-HT_{1A}, 5-HT_{1B}, 5-HT_{1D}, 5-ht_{1E}, 5-ht_{1F}, 5-HT_{2A}, 5-HT_{2B}, 5-HT_{2C}, 5-HT3, 5-HT_4, 5-ht_{5A}, 5-HT_{5B}, 5-ht_6, 5-HT_7	hydroxytryptamine receptors
M_1-M_5	acetylcholine receptors (muscarinic)
MEL_{1A}, MEL_{1B}	melatonin receptors
$mglu_1$-$mglu_8$	metabotropic receptors for excitatory amino acids
NK_1-NK_3	tachykinin receptors
NMDA	glutamate and glycine receptors
PAR_1, PAR_2	protease-activated receptors
$P2X_1$-$P2X_7$	purinoceptor channels
$P2Y_1$, $P2Y_2$, PSY_4, $P2Y_6$	G-protein-coupled receptors
sst_1-sst_5	somatostatin receptors
V_{1A}, V_{1B}, V_2, OT	vasopressin and oxytocin receptors
VIP_1, VIP_2	vasoactive intestinal peptide receptors

Y_1-Y_5, y_6	neuropeptide Y receptors
μ, δ, κ, ORL1	opioid and opioid-like receptors

The same holds for some voltage-gated channels, such as the following:

K_M, K_{ACh}, BK_{Ca}, IK_{Ca}, K_A, K_V, $K_{V(r)}$, $K_{V(s)}$, K_{SR}	K^+ channels
L, N, P, Q, T	Ca^{2+} channels
I, II, III, μ1, H1, PN3	Na^+ channels

Immunology

A, B, 0 (zero, not O)	blood group antigens of glycoprotein and muco-polysaccharide nature
Ab, ab	antibody
ABC	antigen-binding cell; antigen-binding capacity
AbFC	Ab-forming cell
ACIF	anticomplement immunofluorescence
ADCC	antibody-dependent cell-mediated cytotoxicity
AFP	α-fetoprotein
Ag, ag	antigen
AIF	anaphylatoxin-inhibitory factor
ANCA	anti-neutrophil cytoplasmic antibody
APC	antigen-presenting cell
ARAM	antigen-recognition activating motif
ARC	antigen-reactive cell
B (lymphocyte *or* cell)	cell formed in the *bursa Fabricii* of birds and bone marrow of mammals
B, D, H, I, P	complement factors in the alternative pathway
BAF	B-cell-activating factor
BCDF	B-cell differentiation factor
BCGF	B-cell growth factor
C, D, J, L, V	segments (or subgenes) of a gene coding for immunoglobulins
C-region	constant-composition part of an immunoglobulin
C_H	constant region of heavy chain
C_L	constant region of light chain
C1q, C1r, C1s, C2, C3a, C3b, C3dg, C4, C5a, C5b, C6, C7, C8, C9	proteins of the mammalian complement system

[A bar over the numerals (e.g., $\overline{C3b}$) designates an active species.]

CA	cytotoxic antibody
CAM	cell-adhesion molecule
CD1a – CDw166 antigens (in December 1998)	differentiation antigens located on cell surfaces (*for* cluster of differentiation)
CCP	complement-control protein
CFA	complete Freund's adjuvant

(continued)

CFR	complement fixation reaction
CFT	complement fixation test
CH/RG	Chido/Rogers blood group system
CIS	central immunity system
CLMF	cytotoxic-lymphocyte maturation factor
CMC	cell-mediated cytotoxicity
CMI	cell-mediated immunity
CMIR	cell-mediated immune response
CML	cell-mediated cytolysis
CO	Colton blood group system
CR	Cromer blood group system
CR1-CR5	complement factor receptor proteins
CRL	complement-receptor-bearing lymphocytes
CSAT	cell-substrate attachment antigen
CTh	carrier-specific T helper cell
CTL	cytotoxic T lymphocyte
CTMC	connective tissue mast cell
D-region	hypervariable region between the V and J regions in Ig heavy chain
DC	dendritic cell
DI	Diego blood gropup system
DID	double immunodiffusion
DO	Dombrock blood group system
DTH	delayed-type hypersensitivity
EID	electroimmunodiffusion
ELFIA	enzyme-linked fluorescence immunoassay
ELIA	enzyme-labeled immunoassay
ELISA	enzyme-linked immunosorbent assay
EMIA	enzyme membrane immunoassay
EMIT	enzyme multiplied immunoassay technique
ENA	extractable nuclear antigen
Eos	eosinophil leukocyte
F antigens	fimbrial antigens
Fab, Fc, Fd, Fd', Fr	fragments of the immunoglobulin molecule
$Fc_\alpha R$	receptor for Fc fragment of heavy chain A
factors B, D, H, I	proteins involved in complement fixation
FCA	Freund's complete adjuvant
FDC	follicular dendritic cell
FGF	fibroblast growth factor
FIA	fluorescence immunoassay
FY	Duffy blood group system

(continued)

GCSF	granulocyte colony-stimulating factor
GE	Gerbich blood group system
GM-CSF	granulocyte-macrophage colony-stimulating factor
GVHR	grant-versus-host reaction (or response)
H antigen	1. bacterial flagellar antigen;
	2. precursor of the A and B blood group antigens
	3. a histocompatibility antigen
	4. Hh blood group system
H chain	heavy chain of immunoglobulin
HLA	human leukocyte antigen
IAC	immunologically activated cell
IBF	immunoglobulin-binding factor
ICC	immunologically competent cell
Id	idiotype
IEL	intraepithelial lymphocyte
IEM	immunoelectron microscopy
IEOP	immunoelectroosmophoresis
IEP	immunoelectrophoresis
Ig (IgA, IgD, IgE, IgG, IgM, IgX)	immunoglobulin (of class A–X)
IIF	indirect immunofluorescence
IFN	interferon

[Interferons of human and mouse origin are designated with Hu and Mo, respectively; the type of interferon is designated with a Greek letter, e.g., HuIFM-α.]

IL1 - IL17	interleukins 1-17
Ir (genes)	immune-response genes
IRIA	indirect radioimmunoassay
J chain	chain present in polymeric IgA and IgM
JK	Kidd blood group system
K antigens	capsular antigens
KEL	Kell blood group system
KN	Knops blood group system
L cell	a cell of a mouse sarcome cell line
L chain	light chain of immunoglobulin
LAF	lymphocyte-activating factor
LAK cells	lymphokine-activated killer cells
LCS	leukocyte complement system
Le	Lewis antigen
Le^a, Le^b	Lewis blood group substances
LE	Lewis blood group system
LFA	lymphopcyte-function-related antigen
LGL	large granular lymphocyte

(continued)

LIF	leukemia inhibitory factor
LT	lymphotoxin
LU	Lutheran blood group system
LW	Landsteiner-Weiner blood group system
M, N, S	blood group antigens associated with glycophorin
mAb	monoclonal antibody
MAC	membrane attack complex
MAF	macrophage-activating factor
MC	memory cell
MCAF	monocyte chemotaxis-activating factor
MCGF	mast-cell growth factor
MCP	macrophage-capping protein; monocyte chemotactic protein
MCSF	macrophage colony-stimulating factor
MDGF	macrophage-derived growth factor
MFF	macrophage fusion factor
MHC	major histocompatibility complex
MIF	macrophage inhibition factor; migration inhibiting factor
MIP	macrophage inflammatory protein
Mono	monocyte
MPI	monocyte production inhibitor
MPS	monocyte-phagocyte system
MΦ	macrophage
NAP	neutrophil-activating protein
NIF	neutrophil-immobilizing factor
NK cells	natural killer cells
PBA	polyclonal B-cell activator
PBL	peripheral blood lymphocyte
PBMC	peripheral blood mononuclear cell
PC	plasma cell
PFC	plaque-forming cell
PME	polymorphonuclear eosinophil
PMN(L)	polymorphonuclear lymphocyte
RFC	rosette-forming cell
Rh (factor)	antigen of the Rh blood group system (*for* Rhesus)
RH	Rh blood group system
RI	radioimmunology
SC	stem cell
SCF	stem cell factor
sIG	surface Ig; soluble Ig; secretory Ig

(continued)

SRID	single radial immunodiffusion
T (lymphocyte *or* cell)	cell formed in the thymus
TAM	tyrosine-based activation motif
Tc (cell)	cytotoxic T lymphocyte
TCGF	T-cell growth factor
TCR	T-cell receptor
TD	thymus-dependent antigen
TF	transfer factor
TGF	transformation growth factor
Th (cell)	helper T lymphocyte
Thy	thymocyte
TI-1, TI-2	thymus-independent antigen
TIL	tumor-infiltrating lymphocyte
TP	thymopoietin
TR	target receptor
Ts (cell)	su ppressor T lymphocyte
VLA	very late antigen
V-region	variable part of an immunoglobulin molecule
XG	Xg blood group system
XK	Xk blood group system
YT	Yt blood group system
ZAP	ζ-associated protein
$\alpha, \delta, \epsilon, \gamma, \mu$	heavy chains of IgA, IgD, IgE, IgG and IgM
κ, λ	light chains of immunoglobulins

Medicine

This is a field where no binding rules for creating abbreviations and acronyms exist. Those that are listed here have been gleaned from a number of medical journals and treatises, as well as from guides and manuals for biology editors. The abbreviations begin with those used in general medical practice and are followed by those used for various diseases and syndromes. An additional section on techniques used in medical diagnosis is included

General

a (also subscript)	arterial (blood)
A	alveolar (gas)
A	atrium
AF	amniotic fluid
ANS	autonomic nervous system
ap	anterior posterior
av	arteriovenous
aw	airway
b, B (also subscript)	blood
BBB	blood-brain barrier
BCG	bacille Calmette-Guérin
BMR	basal metabolic rate (W kg^{-1})
BOB	blood-ocular barrier
BP	blood pressure
BS	blood sugar
BX	biopsy
c	capillary (blood)
C	content of gas per volume of blood
C_f	capillary filtration factor
C_m	clearance factor
CA	cardiac arrest
CAM	chorio-allantoic membrane
CBF	cerebral blood flow; coronary blood flow
CC	contractile component
CFU	colony-forming unit

(continued)

CL_R	renal clearance of drug
CL_{NR}	nonrenal clearance of drug
CL_{CR}	creatinine clearance
CNS	central nervous system
CO, Q_{LV}	cardiac output (L min^{-1})
COD	chemical oxygen demand
CPE	cytopathogenic effect
CR	conditioned reflex
CSF	cerebrospinal fluid
CVP	central venous pressure
D	dead-space (gas)
DAT	differential agglutination test
diast (also subscript)	diastolic
e (also subscript)	excreted into urine
E	expired (gas)
E_H	hepatic extraction ratio
ECF	extracellular fluid
ECM	extracellular matrix
H (also subscript)	hepatic
H	metabolic heat production (W m^{-2})
HFV	high-grequency ventilation
i	ideal
I	inspired (gas)
i.c., ic	intracutaneous(ly)
i.d., id	intradermal(ly)
i.g., ig	intragastrical(ly)
i.m., im	intramuscular(ly)
i.p., ip	intraperitoenal(ly)
i.v., iv	intravenous(ly)
ISF	interstitial fluid
L	lung
Lym	lymphocytic
m (also subscript)	alveolar capillary membrane; metabolized
m, M (as prefixes)	mini-, maxi-; micro-, macro-; minor, major
MCR	metabolic clearance rate
MEP	miniature endplate potential
MLD	minimum lethal dose
Myl	myeloid
Neu	neural
p (also subscript)	plasma
P	pulmonary

(continued)

P	pressure (in kPa; in clinical use still in mm Hg)
p.o., po	per os (orally)
p.r., pr	per rectum
Q	volume quantity (L)
R	resistance (kPa L^{-1})
RBC	red blood cell
RQ	respiratory exchange ratio
s	shunt
s.c., sc	subcutaneous(ly)
s.l., sl	sublingual(ly)
sal	saliva
SNS	sympathetic nervous system
syst	systolic
t_{max}	time to reach maximum concentration of drug after administration
t (also subscript)	tissue
th (also subscript)	thorax
top (also subscript)	topical
US	unconditioned stimulus
v, ven (also subscript)	venous (blood)
V	ventricle
va	venous admixture
w (also subscript)	water
γ	absolute humidity (kg m^{-3})

Diseases

AA	agranulocytic angina; aplastic anemia
AAA	acquired aplastic anemia
AAE	acute allergic encephalitis
AASP	acute atrophic spinal paralysis
AB	Alibert-Bazin (disease); asthmatic bronchitis
ABA	Addison-Biermer anemia
ABE	acute bacterial endocarditis
ACA	adenocarcinoma
ACC	alveolar cell carcinoma; articular chondrocalcinosis
ACD	allergic contact dermatitis; adult celiac disease
ACED	anhydrotic congenital ectodermal dysplasia
ACH	active chronic hepatitis
ACI	acute coronary infarction; acute coronary insufficiency
AD	Aleutian disease; Alzheimer's disease; atopic dermatitis
ADE	acute disseminated encephalitis

(continued)

ADEM	acute disseminated encephalomyelitis
AEB	acute erythroblastopenia
AEN	aseptic epiphyseal necrosis
AER	acute exertional rhabdomyolysis
AFRD	acute febrile respiratory disease
AG	Addison-Gull (disease)
AGG	agammaglobulinemia
AGL	acute granulocytic leukemia
AGN	acute glomerulonephritis
AGS	adrenal gland sympathogonioma
AGT	acute generalized tuberculosis
AH	acute hepatitis; astigmatic hypermetropy
AHA	acute hemolytic anemia; autoimmune hemolytic anemia
AHC	acute hemorrhagic conjunctivitis; acute hemorrhagic cystitis
AHD	atherosclerotic heart disease; autoimmune hemolytic disease
AHE	acute hemorrhagic encephalomyelitis
AHLE	acute hemorrhagic leukoencephalitis
AHO	Albright's hereditary osteodystrophy
AHP	acute hemorrhagic pancreatitis
AID	autoimmune disease
AIE	acute infectious encephalitis
AIHA	autoimmune hemolytic anemia
AIL	acute infectious lymphocytosis
AIP	acute infectious polyneuritis; acute intermittent porphyria
AL	acute leukemia
ALD	adrenoleukodystrophy; alcoholic liver disease
ALL	acute lymphoblastic leukemia
ALS	amyotrophic lateral sclerosis
ALTB	acute laryngotracheobronchitis
AMML	acute monomyelocytic leukemia
AMT	acute miliary tuberculosis
AN	aseptic necrosis; avascular necrosis
AND	algoneurodystrophy
ANLL	acute nonlymphocytic leukemia
ANUG	acute necrotizing ulcerative gingivitis
AOD	arterial occlusive disease
APCKD	adult polycystic kidney disease
APD	autoimmune progesterone dermatitis
APE	acute polioencephalitis
APSGN	acute poststreptococcal glomerulonephritis
AR	atrophic rhinitis

(continued)

ARD	acute respiratory disease
AS	Adam-Stokes (disease); ankylosing spondylitis; atherosclerosis
ASH	asymmetrical septal hypertrophy
ASHD	arteriosclerotic heart disease
ASN	arteriosclerotic nephritis
ATD	asphyxiating thoracic dystrophy
ATL	adult T-cell leukemia
ATM	acute transverse myelopathy
ATN	acute tubular necrosis
AuSH	Australian serum hepatitis
AVD	aortic valvular disease
AVH	acute viral hepatitis
AYA	acute yellow atrophy
BA	bronchial asthma
BANF	bilateral acoustic neurofibromatosis
BB	Besnier-Bock (disease); Bogdän-Buday (disease); Busse-Buschke (disease)
BCC	basal cell carcinoma
BCE	basal cell epithelioma
BCH	basic cell hyperplasia
BG	Beurmann-Gougerot (disease)
BHF	Bolivian hemorrhagic fever
BJ	Bielschowsky-Jansky (disease)
BL	Burkitt's lymphoma
BM	Bamberger-Marie (disease)
BMD	Becker muscular dystrophy
BME	biundulant meningoencephalitis
BMN	bone marrow necrosis
BPH	benign prostatic hyperplasia
BS	Baehr-Schiffrin (disease); Brill-Symmers (disease); Brown-Séquard (disease); Brown-Symmers (disease)
BWD	bacillary white diarrhea
CAHD	coronary arteriosclerotic heart disease
CASMD	congenital atonic sclerotic muscular dystrophy
CD	cardiovascular disease; celiac disease; Crohn¢s disease
CDA	congenital dyserythropoietic anemia
CDD	chronic degenerative disease
CE	California encephalitis; Charcot-Erb (disease)
CETE	Central European tick-borne encephalitis
CF	cystic fibrosis
CGD	chronic granulomatous disease

(continued)

CGN	chronic glomerulonephritis
CHD	congenital heart disease; congestive heart disease; coronary heart disease; cyanotic heart disease
CHF	congenital hepatic fibrosis; congestive heart failure
CHH	cartilage-hair hypoplasia
CHLD	chronic hypoxic lung disease
CHMD	clinical hyaline membrane disease
CIBHA	congenital inclusion-body hemolytic anemia
CID	cytomegalic inclusion disease; combined immunodeficiency disease
CIN	cervical intraepithelial neoplasia
CJD	Creutzfeldt-Jakob disease
CLAH	congenital lipoid adrenal hyperplasia
CLE	centrilobular emphysema
CLL	chronic lymphocytic leukemia
CLT	chronic lymphocytic thyroiditis
CM	cardiomyopathy
CML	cell-mediated lympholysis;chronic myelogenous leukemia
CMPGN	chronic membranoproliferative glomerulonephritis
CN	congenital nephrosis; congenital nystagmus; Crigler-Najjar (disease)
CNSHA	congenital nonspherocytic hemolytic anemia
COLD	chronic obstructive lung disease
COPD	chronic obstructive pulmonary disease
CPEO	chronic progressive external ophthalmoplegia
CPGN	chronic proliferative glomerulonephritis
CPL	congenital pulmonary lymphangiectasis
CPM	central pontine myelinolysis
CPN	chronic polyneuropathy; chronic pyelonephritis
CPS	contagious pustular stomatitis
CR	Cacchi-Ricci (disease)
CRD	chronic renal disease; chronic respiratory disease
CS	chronic schizophrenia
CSM	cerebrospinal meningitis
CTD	connective tissue disease
CUC	chronic ulcerative colitis
CVA	chronic villous arthritis
CVD	cardiovascular disease; cerebrovascular disease
CVH	common variable hypogammaglobulinemia
CVID	common variable immunodeficiency
CW	Christian-Weber (disease)
DA	degenerative arthritis
DB	Diamond-Blackfan (disease)

(continued)

DCS	diffuse cerebral sclerosis
DD	Di Guglielmo's disease; Dupuytren's disease
DF	Daae-Finsen (disease)
DG	Duchenne-Griesinger (disease)
DGN	diffuse glomerulonephritis
DGS	diabetic glomerulonephritis
DH	disseminated histoplasmosis
DHF	Dengue hemorrhagic fever
DIF	diffuse interstitial fibrosis
DIP	desquamative interstitial pneumonia
DJD	degenerative joint disease
DMD	Duchenne muscular dystrophy
DME	diphasic meningoencephalitis
DNF	Durand-Nicolas-Favre (disease)
DR	diabetic retinopathy
DSAP	disseminated superficial actinic porokeratosis
DT	delirium tremens
DVD	double vessel disease
DVT	deep vein thrombosis
EACD	eczematous allergic contact dermatitis
EECD	endothelial-epithelial corneal dystrophy
EFE	endocardial fibroelastosis
EG	Erb-Goldflam (disease)
EHF	epidemic hemorrhagic fever
EKC	epidemic keratoconjunctivitis
EMC	encephalomyocarditis
EMF	endomyocardial fibrosis
EP	erythropoietic protoporphyria
FA	Fanconi's anemia
FAP	familial amyloidotic polyneuropathy
FCD	fibrocystic dysplasia
FCHL	familial combined hyperlipidemia
FD	Filatow-Dukes (disease)
FETE	Far Eastern tick-borne encephalitis
FF	Fox-Fordyce (disease)
FGD	fatal granulomatous disease
FGN	focal glomerulonephritis
FH	familial hypercholetsreolemia; fibromuscular hyperplasia
FHA	familial hypoplastic anemia
FHC	familial hypertrophic cardiomyopathy
FMD	fibromuscular dysplasia; foot-and mouth disease

(continued)

FMF	familial Mediterranean fever
FMH	fibromuscular hyperplasia
FNC	fatty nutritional cirrhosis
FNH	focal nodular hyperplasia
FS	fibrosarcoma
FSGN	focal sclerosing glomerulonephritis
FV	Fahr-Volhard (disease)
GERD	gastrophageal reflux disease
GH	Gee-Herter (disease); Gilford-Hutchinson (disease)
GIP	gonorrheal invasive peritonitis
GN	Gandy-Nanta (disease)
GT	Gee-Thaysen (disease)
GTN	glomerulotubulonephritis
HB	hepatitis B; Hutchinson-Boeck (disease)
HC	Huntington's chorea
HCC	hepatocellular carcinoma; heavy-chain disease
HCL	hairy-cell leukemia
HCM	hypertrophic cardiomyopathy
HCP	hereditary coproporphyria
HD	Hansen's disease; Hirschsprung's disease; Hodgkin's disease; Huntington's disease
HE	hemagglutinating encephalomyelitis; hereditary elliptocytosis
HG	Hutchinson-Gilford (disease)
HIT	hypertrophic infiltrative tendonitis
HJ	Hebra-Jadassohn (disease)
HM	Hebra-Medin (disease)
HMD	hyaline membrane disease
HOCM	hypertrophic obstructive cardiac myopathy
HOS	human osteosarcoma
HS	hereditary spherocytosis; herpes simplex; Holländer-Simons (disease)
HSC	Hand-Schüller-Christian (disease)
HT	Hashimoto's thyroiditis
HW	His-Werner (disease)
IAP	intermittent acute porphyria
IB	Ibrahim-Beck (disease)
IBD	inflammatory bowel disease
ICD	ischemic coronary disease
IDF	infantile digital fibromatosis
IDDM	insulin-dependent diabetes mellitus
IEC	intraepithelial carcinoma
IGA	infantile genetic agranulocytosis

(continued)

IHC	idiopathic hemochromatosis
IHD	ischemic heart disease
IHH	infectious human hepatitis
IJD	inflammatory joint disease
IPE	interstitial pulmonary emphysema
IPH	idiopathic pulmonary hemosiderosis; inflammatory papillary hyperplasia
IRF	idiopathic retroperitoneal fibrosis
JBE	Japanese B encephalitis
JCP	juvenile chronic polyarthritis
JGP	juvenile general paralysis
JL	Jaksch-Luzet (disease)
JRA	juvenile rheumatoid arthritis
KB	Kaschin-Beck (disease)
KD	Kawasaki disease
KHD	kinky hair disease
KHS	Korean hemorrhagic fever
KM	Kraepelin-Morel (disease)
KS	Kaposi's sarcoma
KW	Kugelberg-Welander (disease)
LA	Lightwood-Albright (disease)
LCM	lymphocytic choriomeningitis
LCP	Legg-Calvé-Perthes (disease)
LD	Legionnaires' disease
LE	lupus erythematosus
LHON	Leber's hereditary optic neuropathy
LJ	Larsen-Johansson (disease)
LL	lymphoid leukemia
LN	lipoid nephrosis
LS	Letterer-Siwe (disease); Libman-Sacks (disease)
LTB	laryngotracheal bronchitis
MA	mammary adenocarcinoma
MAT	multifocal atrial tachycardia
MB	Marie-Bamberger (disease)
MCG	mesangiocapillary glomerulonephritis
MD	muscular dystrophy
MEA	multiple endocrine adenomatosis
MELAS	mitochondrial encephalomyopathy, lactic acidosis, and stroke-like episodes
MERRF	myoclonic epilepsy with ragged red fibers
MFD	mandibulofacial dysostosis

(continued)

MG	membranous glomerulonephritis; myasthenia gravis
MGP	membranous glomerulopathy
MHA	microangiopathic hemolytic anemia
MI	myocardial infarction
MK	More-Kraepelin (disease)
ML	malignant lymphoma
MMD	myotonic muscular dystrophy
MN	malignant nephrosclerosis
MP	Merzbacher-Pelizaeus (disease)
MS	Marie-Strümpell (disease); multiple sclerosis; mitral stenosis;
MSL	Murphy-Sturm lymphosarcoma
NANB	non-A-non-B (hepatitis)
ND	neoplastic disease; Newcastle disease
NDF	Nicolas-Durand-Favre (disease)
NEC	necrotizing enterocolitis
NGU	nongonococcal urethritis
NMM	nodular malignant melanoma
NMTD	nonmetastatic trophoblastic disease
NPD	Niemann-Pick disease
NUG	necrotizing ulcerative gingivitis
NW	Norman-Wood (disease)
OD	occupational disease
OS	Osgood-Schlatter (disease)
OU	Oppenheim-Urbach (disease)
OV	Osler-Vaquez (disease)
PA	pernicious anemia
PAME	primary amebic meningoencephalitis
PBC	primary biliary cirrhosis
PC	portal cirrhosis; prostatic carcinoma
PCD	primary ciliary dyskinesis
PCF	pharyngoconjunctival fever
PE	Pel-Ebstein (disease)
PFD	polyostotic fibrous dysplasia
PID	paralytic infantile paralysis; pelvic inflammatory disease
PKD	proliferative kidney disease
PKU	phenylketonuria
PM	papular mucinosis
PMD	primary myocardial disease; progressive muscular dystrophy
PML	progressive multifocal leukodystrophy; progressive multifocal leukoencephalopathy promyelocytic leukemia
PN	polyneuritis; pyelonephritis

(continued)

PS	Paltauf-Sternberg (disease); pulmonary stenosis
PSA	progressive spinal ataxia
PSS	progressive systemic sclerosis
PVD	peripheral vascular disease
RA	refractory anemia; rheumatoid arthritis
RAS	recurrent aphthous stomatitis
REM	reticular erythematous mucinosis
RF	rheumatic fever
RM	Rothmann-Makai (disease)
RMSF	Rocky Mountain spotted fever
ROW	Rendu-Osler-Weber (disease)
RSSE	Russian spring-summer encephalitis
RTA	renal tubular acidosis
RTCS	Roske-De Toni-Caffey-Smith (disease)
RW	Romano-Ward (disease)
SA	Stokes-Adams (disease)
SBE	subacute bacterial endocarditis
SC	Schüller-Christian (disease)
SCA	sickle-cell anemia
SCID	(congenital) severe combined immunodeficiency
SJ	Stevens-Johnson (disease)
SL	Sinding-Larsen (disease); Strümpell-Lorain (disease)
SLE	Saint Louis encephalitis; systemic lupus erythematosus
SMA	spinal muscular atrophy
SO	Schlatter-Osgood (disease)
SPA	spinal progressive amyotrophy
SPK	superficial punctate keratitis
SPMA	spinal progressive muscular atrophy
SS	Spillinger-Stock (disease)
SSM	superficial spreading melanoma
SSPE	subacute sclerosing panencephalitis
SU	Senear-Usher (disease)
SV	Spielmeyer-Vogt (disease)
SVD	single-vessel disease
SW	Sturge-Weber (disease)
TAD	transient acantholytic dermatosis
TB	tuberculosis
TBE	tick-borne encephalitis
TD	tibial dyschondroplasia; typhoid dysentery
TEN	toxic epidermal necrosis
TSD	Tay-Sachs disease

(continued)

UD	underline{u}lcerative underline{d}ermatosis
UDN	underline{u}lcerative underline{d}ermal underline{n}ecrosis
UPI	underline{u}teroplacental underline{i}schemia
URVD	underline{u}nilateral renounderline{v}ascular underline{d}isease
VD	underline{v}enereal underline{d}isease; underline{v}essel underline{d}isease
VHD	underline{v}alvular underline{h}eart underline{d}isease; underline{v}entral underline{h}eart disease
VOD	underline{v}enous underline{o}cclusive underline{d}isease
VS	underline{v}esicular underline{s}tomatitis
VWD	underline{v}on underline{W}illebrand underline{d}isease
WD	underline{W}ilson¢s underline{d}isease
WK	underline{W}ilson-underline{K}immelstiel (disease)
XDP	underline{x}eroderma underline{p}igmentosum
XGP	underline{x}anthogranulomatous underline{p}yelonephritis
XID	underline{X}-linked underline{i}mmunounderline{d}eficiency
XLMR	underline{X}-underline{l}inked underline{m}ental underline{r}etardation
YF	underline{y}ellow underline{f}ever

Syndromes

If the syndrome is named after more than one discoverer, its abbreviation does not contain the letter S.

ABB	Albright-Buttler-Bloomberg
ABL	Albright-Buttler-Lightwood
AC	Arnold-Chiari
AD	Adair-Dighton
ADS	antibody deficiency s.
AH	Adie-Holmes
AIDS	underline{a}cquired underline{i}mmunoounderline{d}eficiency underline{s.}
AL	Albright-Lightwood
AR	Arnold-Realy
ARDS	adult respiratory distress s.
AS	Angelman s.
AT	Achard-Thiers
BC	Budd-Chiari
BCS	battered child s.
BD	Blackfan-Diamond; blue diaper s.
BF	burning feet s.
BG	Barré-Guillain; Bertolotti-Barcin; Brock-Graham; Buerger-Grütz
BH	Bernard-Horner; Bogaert-Hozay
BK	Bassen-Kornzweig
BL	Barré-Liéou

(continued)

BM	Batten-Mayou; Bingold's myorenal s.; Brailsford-Morquio
BN	Babinski-Nageotte
BP	Bard-Pic
BS	Batten-Steinert; Bloch-Sulzberger; Bloom's; Boyd-Stearns
BSV	Batten-Spielomeyer-Vogt
BT	Behçet-Touraine
BTS	bradytachy s.
BU	Bonnevie-Ullrich
BW	Beckwith-Wiedermann
BWG	Bland-White-Garland
CB	Cruveilhier-von Baumgarten
CBS	Curschmann-Batten-Steinert
CC	Cestan-Chenais
CE	Camurati-Engelmann
CH	Chédiak-Higashi; Clarke-Hadfield; Conradi-Hünermann
CHRS	cerebrohepatorenal s.
CMT	Charcot-Marie-Tooth
COBS	chronic organic brain s.
CPA	cerebellopontine angle s.
CRS	Chinese restaurant s.; congenital rubella s.
CS	Cockayne's s.; Collet-Sicard; Curschmann-Steinert
CST	Christ-Siemens-Touraine
CT	Courvoisier-Terrier
CW	Cavare-Westphal
CWB	Charcot-Weiss-Baker
DB	Dollinger-Bielschowsky
DDT	Degos-Delort-Tricot
DE	Duchenne-Erb
DF	Debré-Finsen
DJ	Dubin-Johnson
DK	Déjerine-Klumpke
DL	Duchenne-Leyden
DR	Déjerine-Roussy
DS	Debré-Semelaigne; Déjerine-Sottas; deprivation s.; Down's s.
DSS	Dengue shock s.
DT	Déjerine-Thomas
DTF	Debré-De Toni-Fanconi
DY	Dyke-Young
DZ	Durand-Zunin
EC	Ellis-van Creveld; Erb-Charcot
ED	Ehlers-Danlos

(continued)

EL	Eaton-Lambert
EPS	extrapyramidal side-effect s.
ES	Eddowes' s.
FA	Forbes-Albright
FAZ	Fanconi-Albertini-Zellweger
FDG	Fibiger-Debré-Gierki
FH	Fanconi-Hegglin
FK	Feil-Klippel; Foster-Kennedy
FL	Fiessinger-Leroy
FLR	Fiessinger-Leroy-Reiter
FMS	Frankfurt-Marburg s.
FP	Fanconi-Petrassi
FR	Favre-Racouchot
FS	Freeman-Sheldon
FT	Fanconi-De Tari
FW	Falconer-Wadell; Friderichsen-Waterhouse
GAR	genitoanorectal s.
GB	Gilbert-Behçet; Gougerot-Blum; Guillain-Barré
GC	Gougerot-Carteuad
GE	Gänsslen-Erb; Gsell-Erdheim
GH	Gougerot-Hailey
GK	Gasser-Karrer
GL	Gilbert-Lereboullet
GMM	Goldberg-Maxwell-Morris
GO	Gordon-Overstreet
GS	Gilbert's; Glanzmann-Saland; Gougerot-Sjögren; Groenblad-Strandberg; Guérin-Stern
HD	Haab-Dimmer
HFRS	hemorrhagic fever with renal s.
HG	Herter-Gee
HH	Hunter-Hurler
HK	Hoffa-Kastert
HM	Hertwig-Magendie
HMB	Horton-Magath-Brown
HO	Holt-Oram
HR	Hamman-Rich; Howship-Romberg
HS	Hallevorden-Spatz
HTS	head traumatic s.
HW	Hayem-Wahl; Hertwig-Weyers
HWP	Hutchinson-Weber-Peutz
IRDS	idiopathic respiratory distress s.

(continued)

JL	Jadassohn-Lewandowski; Jaffe-Lichtenstein
KK	Kaplan-Klaskin
KL	Kleine-Levin
KRA	Klinefelter-Reifenstein-Albright
KS	Kugel-Stoloff
KT	Klippel-Trenaunay
KW	Kimmelstiel-Wilson
KZ	Kaplan-Zuelzer
LB	Lederer-Brill
LC	Lannois-Cleret
LD	Lucey-Driscoll
LEMS	Lambert Eaton myasthenic s.
LGB	Landry-Guillain-Barré
LK	Landry-Kussmaul; Löhr-Kindberg
LL	Lewandowsky-Lutz
LM	Looser-Milkma
LMB	Laurence-Moon-Biedl
LN	Lesch-Nyhan
LS	Luciani s.
LW	Léri-Weill
MA	Martin-Albright
MAU	Meyenburg-Altherr-Hehlinger
MCG	Minkowski-Chauffard-Gänsslen
MCLS	mucocutaneous lymph node s.
MDS	multiple deficiency s.
MED	Meckeren-Ehlers-Danlos
MF	Miller-Fischer
MG	Millard-Gubler
MK	Mounier-Kuhn
ML	Marie-Léri
MLS	middle lobe s.
MM	Morel-Morgagni
MO	Minor-Oppenheim
MR	Melkersson-Rosenthal
MRC	Müller-Ribbing-Clement
MS	Marfan's s.; Meckel's s.
MvW	Minot-von Willebrand
MW	Mallory-Weiss
NL	Nyhan-Lesch
NMM	Nonne-Milroy-Meige
NS	nephrotic s.

(continued)

OBS	organic brain syndrome
OF	Ostrum-Furst
OFD	oral-facial-digital s.
OLA	Ollier-Klippel-Trenaunay
OMS	organic mental syndrome
OT	Oesterreicher-Turner
PCOS	polycystic ovarian s.
PHM	pulmonary hyaline membrane s.
PJ	Peutz-Jeghers
PK	Parrot-Kaufmann; Piringer-Kuchinka
PMTS	premenstrual tension s.
PS	Paget-von Schroetter
PV	Plummer-Vinson
PW	Prader-Willi
PWC	Pfeifer-Weber-Christian
RB	Roth-Bernhardt
RC	Raymond-Cestan; Roussy-Cornil
RD	Renon-Delile; Riley-Day
RDS	respiratory distress s.
RF	Riga-Fede; Rundles-Falls
RGM	Rietti-Greppi-Micheli
RH	Richner-Hanhart
RL	Roussy-Lévy
RM	Rosenthal-Melkerson
RS	Reye's s.
RT	Rubinstein-Taybi
RW	Romano-Ward
SB	Schwartz-Bartter
SD	Shy-Drager
SF	Schiller-Foix; Still-Felty
SG	Strandberg-Groenblad
SH	Schoenlein-Henoch
SIDS	sudden infant death s.
SKW	Sturge-Kalischer-Weber
SL	Sjögren-Larsson; Stein-Leventhal
SM	Scheutauer-Marie
SMS	Scheutauer-Marie-Sainton; stiff man s.
SR	Stransky-Regala
SS	Simmonds-Sheehan; Sjögren's s.; Strachan-Scott
SSS	sick sinus s.

(continued)

SSV	Stock-Spielmeyer-Vogt
ST	Spanlang-Tappeiner; Stewart-Treves
SWS	steroid wasting s.
TAC	thyroid-adrenocortical s.
TB	Taussig-Bing
TFS	testicular feminization s.
THC	Troisier-Hanot-Chauffard
TJ	Troell-Junet
TK	Turner-Kieser
TMJ	temporomandibular joint s.
TSG	Touraine-Solente-Golé
TSS	toxic shock s.
TW	Thibi_rge-Weissenbach
UF	Ullrich-Feichtiger
UT	Ullrich-Turner; Unna-Thost
UW	Urbach-Wiethe
VK	Vogt-Koyanagi
VM	Verner-Morrison
VS	Vogt-Spielmeyer
WA	Widal-Abrani; Wiskott-Aldrich
WC	Weber-Christian
WF	Waterhouse-Friderichsen
WH	Werdnig-Hoffman
WHS	Wolf-Hirschhorn
WM	Wilson-Mikity
WO	Weber-Osler
WPW	Wolff-Parkinson-White
WS	Westphal-Strümpell
ZE	Zollinger-Ellison
ZEC	Zinsser-Engman-Cole
ZK	Zuelzer-Kaplan
ZO	Ziehen-Oppenheim

Techniques

ACG	angiocardiography
BEI	back-scattered electron imaging
CAA	carotid audiofrequency analysis
CAT	computed axial tomography
CEEG	computer-analyzed EEG
CKG	cardiokymography
CSE	cross-sectional echocardiography

(continued)

CT	computed tomography
CVG	contrast ventriculography
DCG	dacryocystography
2-DE	two-dimensional echocardiography
ECAT	emission computed axial tomography
ECG	electrocardiography
ECHO	echocardiography
ECoG	electrocorticography
ECT	emission computed tomography
EDA	electrodermal audiometry
EEG	electroencephalography
EEGA	electroencephalographic audiometry
EESG	evoked electrospinography
EKY	electrokymography
EMG	electromyography
ENG	electronystagmography
EOG	electrooculography
ERA	electroencephalic response audiometry
ERCP	endoscopic retrograde cholangiopancreatography
ERG	electron radiography
ESA	electrosurgical arthroscopy
HECT	head-equivalent computed tomography
PSG	radioallergosorbent test
RNA	radionuclide angiography
SECG	stress electrocardiography
SVA	selective visceral angiography
UCG	ultrasonic cardiography
USEG	ultrasound echography

Soil Biology

Soil science is a multidisciplinary field where biosciences represent but a minor part. Therefore, the selection of abbreviations listed here is far from complete and, in fact, is only marginally associated with biology. Many of the relevant abbreviations and acronyms will be found under Microbiology and Plant Physiology, for example.

Soil Phases

N	anthraquic		PH	phreatic
DU	duripan		PL	placic
FR	fragipan		RU	rudic
GE	gelundic		SA	salic
GI	gilgai		SK	skeletic
IN	inundic		SO	sodic
LI	lithic		TK	takyric
PF	petroferric		YR	yermic

Soil Moisture Regime (listed according to increasing aridity)

AQ	aquic		PU	perudic
PQ	peraquic		US	ustic
AR	aridic		XE	xeric
UD	udic		TO	torric

Soil Temperature Regime (listed according to increasing temperature)

PG	pergelic		HT	hyperthermic
CR	cryic		IF	isofrigid
FR	frigid		IM	isomesic
ME	mesic		IT	isothermic
TH	thermic		IH	isohyperthermic

Vegetation

N	no vegetation		W	woodland
G	grassland		S	shrubland
F	forest		SA	savanna

Soil Color

WH	white	YE	yellow
RE	red	RY	reddish yellow
RS	reddish	GE	greenish
YR	yellowish red	GR	grey
BR	brown	GS	greyish
BS	brownish	BU	blue
RB	reddish brown	BB	bluish-black
YB	yellowish brown	BL	black

Soil Texture

C	clay	CSL	coarse sandy loam
L	loam	LS	loamy sand
CL	clay loam	LVFS	loamy very fine sand
Si	silt	LFS	loamy fine sand
SiC	silty clay	LCS	loamy coarse sand
SiCL	silty clay loam	VFS	very fine sand
SiL	silt loam	FS	fine sand
SC	sandy clay	MS	medium sand
SCL	sandy clay loam	CS	coarse sand
SL	sandy loam	US	sand, unsorted
FSL	fine sandy loam	S	sand, unspecified

Human-Introduced Crops

BA	barley	PO	potato
BE	beans	RI	rice
CA	cassava	RB	rice (flooded)
CC	coconut	RU	rice (upland)
CF	coffee	RR	rubber
CH	cashew	SB	soybean
CO	cocoa	SC	sugar cane
CP	cowpea	SF	sunflower
CT	cotton	SO	sorghum
FR	fruit trees	SP	sweet potato
GR	groundnut	TB	tobacco
MA	maize	TE	tea
MI	millet	VE	vegetables
OP	oilpalm	WH	wheat
PE	peas	YA	yams

Master Soil Horizons

H, O, A, E, B, C, R

Paleontology

Symbols for the various geological periods are given.

Quaternary	Q	Mississippian	M
Tertiary	T	Carboniferous	C
Cretaceous	K	Devonian	D
Jurassic	J	Silurian	S
Triassic	T_R	Ordovician	O
Permian	P	Cambrian	$\in$
Pennsylvanian	IP	(Precambrian	$p\in$)

Psychology

Units Uniquely Used

decibel (dB)	*see in physics*
decile	one of the 9 points that divide a ranked distribution into 10 divisions
mel	unit of pitch on a psychophysical scale
sone	unit of loudness on a psychophysical scale
stanine	standard score

Acronyms For Tests, Examinations, Etc.

ACT	American College Testing (assessment)
AGCT	Army General Classification Test
BDI	Beck Depression Inventory
CAT	Children's Apperception Test
CMMS	Columbia Mental Maturity Test
CPI	California Personality Inventory
CPQ	California Personality Questionnaire
CTMM	California Test of Mental Maturity
DAP	Draw-a-Person (test) (= Machover test)
DAT	Differential Aptitude Tests
DIS	Diagnostic Interview Schedule
DSM-III	Diagnostic and Statistical Manual of the American Psychiatric Association
EPI	Eysenck Personality Inventory
EPPS	Edwards Personal Preference Schedule
ETS	Educational Testing Service
FACT	Flanagan Aptitude Classification Test
GATB	General Aptitude Test Battery
GRE	Graduate Record Examinations
KPR	Kuder Preference Record
MAS	Manifest Anxiety Scale
MAT	Miller Analogies Test
MMPI	Minnesota Multiphasic Personality Inventory
MPI	Maudsley Personality Inventory
MPQ	Minnesota Personality Questionnaire

(continued)

PARI	Parental Attitude Research Instrument
PMA	Primary Mental Abilities Test
POI	Personal Orientation Inventory
Q-sort	Forced-Choice Classification of a Fixed Number of Items
SAT	Scholastic Aptitude Test
S-B	Stanford-Binet Test of Intelligence
S-CVII	Strong-Campbell Vocational Interest Inventory
SVIB	Strong Vocational Interest Blank
SVII	Strong Vocational Interest Inventory
TAT	Thematic Apperception Test
TPT	Tactual Performance Test
VIB	Vocational Interest Blank (= Strong Vocational Interest Inventory)
WAIS	Wechsler Adult Intelligence Test
WAT	Word Association Test
W-B	Wechsler-Bellevue Scales
WGTA	Wisconsin General Test Apparatus
WISC	Wechsler Intelligence Scale for Children
WMS	Wechsler Memory Scale
16-D	Bayley Scale
16-P-F	16-Personality-Factors Test

Acronyms For Terms

ADHD	attention-deficit hyperactivity disorder
AEP	auditory evoked potentials
ARMAX	auto-regressive moving averages with exogenous variables (= VARMA)
ASL	American sign language
BARS	behaviorally anchored rating scales
CA	chronological age
CAR	conditioned avoidance response
CE	constant error
CER	conditioned emotional response
cff	critical fusion frequency in vision
CI	conditioned inhibition
CNV	contingent negative variation (in electroencephalography)
CR	Pavlovian (or classically) conditioned response
CRF	schedule of continuous reinforcement for every response
CS	conditioned stimulus
CVC	consonant-vowel-consonant trigram
D	drive
d$'$	detectability or observer's sensitivity
DL	difference limen

(continued)

DMTS	<u>d</u>elayed <u>m</u>atching <u>t</u>o <u>s</u>ample
DOT	<u>d</u>ictionary <u>o</u>f <u>o</u>ccupational <u>t</u>itles
DRL	<u>d</u>ifferential <u>r</u>einforcement of <u>l</u>ow rates of responding schedule of reinforcement
DRO	<u>d</u>ifferential <u>r</u>einforcement of <u>o</u>ther behavior schedule of reinforcement
EDA	<u>e</u>lectro<u>d</u>ermal <u>a</u>ctivity
EDR	<u>e</u>lectro<u>d</u>ermal <u>r</u>esponse
EP	<u>e</u>voked brain <u>p</u>otentials in response to stimulus
EPAM	<u>e</u>lementary <u>p</u>erceiver <u>a</u>nd <u>m</u>emorizer program
EQ	<u>e</u>ducational <u>q</u>uotient
ERIC	<u>e</u>ducational <u>r</u>esources <u>i</u>nformation <u>c</u>enter
ESB	<u>e</u>lectrical <u>s</u>timulation of the <u>b</u>rain
ESP	<u>e</u>xtra<u>s</u>ensory <u>p</u>erception
EU	<u>e</u>xpected <u>u</u>tility
EXT	<u>ext</u>inction (in a schedule of reinforcement)
FFF	<u>f</u>licker <u>f</u>usion <u>f</u>requency
FI	<u>f</u>ixed <u>i</u>ntervals schedule of reinforcements
FR	<u>f</u>ixed <u>r</u>atio schedule of reinforcements
FT	<u>f</u>ixed <u>t</u>ime schedule for delivery of response
g	<u>g</u>eneral intelligence factor *sensu* Spearman
GPS	<u>g</u>eneral <u>p</u>roblem <u>s</u>olver program (Newell and Simon)
GSR	<u>g</u>alvanic <u>s</u>kin <u>r</u>esponse
H	information or uncertainty (in bits) derived from communication/ information theory for a set of events; <u>h</u> for a single event
HAM	early computer model of <u>h</u>uman <u>a</u>ssociative <u>m</u>emory
ICD	<u>i</u>nternational <u>c</u>lassification of <u>d</u>iseases, usually followed by a number (e.g., -9), indicating revision number
ICS	<u>i</u>ntra<u>c</u>ranial <u>s</u>timulation or intracranial self-stimulation
ID	<u>i</u>ndividual <u>d</u>ifference
IQ	<u>i</u>ntelligence <u>q</u>uotient
ITI	<u>i</u>nter-<u>t</u>rial <u>i</u>nterval
JND, jnd	<u>j</u>ust <u>n</u>oticeable <u>d</u>ifference (unit of sensory measure)
K	incentive motivation
L-H	<u>l</u>earned <u>h</u>elplessness
LISREL	popular cross-sectional structural equation modeling program
LTM	<u>l</u>ong-<u>t</u>erm <u>m</u>emory
LTP	<u>l</u>ong-<u>t</u>erm <u>p</u>otentiation
MA	<u>m</u>ental <u>a</u>ge
MTS	<u>m</u>atching <u>t</u>o sample
nAch	<u>n</u>eed <u>ach</u>ievement motivation
nAffil	<u>n</u>eed for <u>affil</u>iation

(continued)

NMTS	non-matching to sample (choose novel); *also* nMTS
OR	orienting response
PA	paired associate (*also* PAL, paired associate learning)
PDP	parallel distributed processing
PERT	program evaluation and review technique
PI	proactive inhibiton
PMA	primary mental abilities
POE	point of objective equality
PRE	partial reinforcement effect
PSE	point of subjective equality
PSI	personalized self-instruction system
PTSD	post-traumatic stress disorder
R	response as in S-R
r_q	fractional antedating goal response
REM	rapid eye movements
RET	rational-emotive therapy
RI	retroactive inhibition
ROC	receiver operating characteristic function
R-S	response-shock interval
RT	reaction time
RTT	"minimal experiment"
R-W theory	popular linear operator model of learning (Rescoria)
S+	stimulus paired with or signaling reinforcement
S-	stimulus signaling absence of reinforcement
SA	social maturity age
SAD	seasonal affective disorder
SDT	signal detection theory
SEC	socio-economic class
S^D	discriminative stimulus, in the presence of which reward is available
SD	differential stimulus, in the presence of which reward is unavailable
S-M	Sado-Masochism
SQ3R	study technique called Survey-Question-Read-Recite-Review
S-R	stimulus-response association or theory
S-S	stimulus-stimulus association interval (*as in* shock-shock interval)
SSDR	species-specific defense reaction (theory)
SST	stimulus sampling theory
stanines	standard scores
STEP	systematic training for effective parenting
STM	short-term memory
T-group	derived from basic skill training groups for interpersonal dynamics; or from sensitivity training groups

(continued)

TE	time error
TO	time-out period, in which reinforcements are not available
TOC, ToC	transfer of control (experimental paradigm)
TOTE	test-operate-test-exit sequence
TRC	truly random control
UCR, UR	unconditioned response
UCS, US	unconditioned stimulus
V	associative strength in several quantitative models of learning
VARIMAX	method for rotating vectors in factor analysis
VEP	visually evoked potentials
VI	variable interval schedule for reinforcements
VTE	vicarious trial and error

Notes On Abbreviations
In General Use
In Scientific Writing

Two tendencies can be observed in contemporary scientific English writing: minimal use of periods (full stops) and a general shift to roman type where italics had been used before.

1. In designating historical periods, use BC or BC for "before Christ," BCE for "before the common era," AD or AD for "anno domini"; with times of the day, use AM for "ante meridiem" and PM for "post meridiem"; likewise, prefer 16:30 to 4.30 PM.

If precise time is to be given, use an analogous notation, e.g., 15:21:40.5 for 15 h, 21 min and 40.5 s.

In abbreviations of academic titles, scientific institutions, etc., use Dr for "Doctor," PhD for "philosophiae doctor," IUBMB for "International Union of Biochemistry and Molecular Biology," and so on.

Many journals now prefer to omit periods in references to literature, both after authors' initials and after journal abbreviations, thus: Baker AC, Dudley EF: ... J Biol Chem ... This may be elegant but there are cases when misunderstanding might arise, such as if the second initial is simply a letter (Harry S Truman!) or if a journal's name contains a word that is not abbreviated.

2. Practically all abbreviations of Latin expressions are now set in roman type. This trend is accompanied by preferring English expressions over Latin ones. A few examples follow.

Abbreviation	Recommended English equivalent
c.,ca.	about, approximately
cf.	compare
e.g.	for example
et al.	and others

(continued)

129

etc.	and so on
ib., ibid.	in the same place
i.e.	that is
infra	below
n.b., NB	take notice
no., nr	number
supra	above
viz.	namely
v., vs.	in contrast to

In addition to the above, a few remarks on style and punctuation are in order.

When referring to a range of numbers, use either the expression "over the range of 1001–2000," or "from 1001 to 2000," not "from 1001–2000."

Dates should be preferably given in the sequence day-month-year, for example, 28 April 1994; alternatively, in the sequence year-month-day, that is 1994-4-28. The commonly used sequence of month-day-year is not recommended in scientific practice.